FLORA OF BERKSHIRE COUNTY

MASSACHUSETTS

To Helen,
A master gardener and
true flower-lover-
Love and best
wishes
Pam

Pamela B. Weatherbee

All proceeds from the sale of this book
will be shared equally by
The Berkshire Museum and the Massachusetts Natural Heritage and
Endangered Species Program

The Berkshire Museum

1996

Cover calligraphy: Ann Kremers

Cover photo: **Rubus odoratus** Purple-flowering raspberry

Printed in the U.S.A.
by

The Studley Press, Inc.
Dalton, Massachusetts

ISBN 0-9648211-0-9

ACKNOWLEDGMENTS

This book is based on a thesis produced as partial fulfillment of the requirements for the degree of Master of Science in botany at the University of New Hampshire. I am deeply indebted to my major professor, Dr. Garrett E. Crow, for accepting my project, and for the help and encouragement that he extended to me. The comments and guidance provided by Dr. Thomas D. Lee and Dr. A. Linn Bogle were also crucial to the completion of the thesis.

A longtime interest in plants was given added incentive by the creation of the Massachusetts Natural Heritage and Endangered Species Program (MNHESP) and their goal of rediscovering and mapping rare and unusual species whose localities had become lost through the intervening years of disinterest. Bruce A. Sorrie, State Botanist from 1979-1992, generously made available his research into old herbarium records and journal papers, which aided immeasureably in the rediscovery of many species in the field. Equally important was his deep knowledge of plants in the field, their characteristics, habitats and communities, which formed the basis of my botanical education.

Many other people provided help along the way. The staff at MNHESP through their database and location maps provided much information about Berkshire County. Their Small Contracts Program, funding searches for many plants, in turn, added to my own knowledge and experience. Ray Angelo, Curator of the New England Botanical Club Herbarium, was of inestimable help in my research into all records for Berkshire County. Friendly landowners allowed me on their land. State personnel at the Department of Environmental Management gave much infomation on trails and features of their lands.

Berkshire naturalists, such as Morgan Bulkely III, Bartlett Hendricks, and Alvah Sanborn, have given clues, hints and information as to hidden bogs, swamps and other rare places. Jerry Jenkins, through his field courses and assembled knowledge, has influenced many discoveries.

Sections on origins of the flora and on plant communities appeared previously in a slightly different form in *Rhodora*, the journal of the New England Botanical Club.

Finally, the all-important task of helping the book into publication has been advanced by David Costello, Director of The Berkshire Museum, who has provided needed guidance in editing and publication procedures. Tracy Wilson has been invaluable in handling promotion and funding for the color plates. Thom Smith lent his his enthusiastic support, and David St. James painstakingly reviewed the Catalog of Vascular Plants section.

THE COLOR PLATES WERE MADE POSSIBLE BY
THE GENEROSITY OF THE FOLLOWING
INDIVIDUALS AND ORGANIZATIONS

Berkshire Natural Resources Council, Inc.

Ursula and Walter Cliff

Bobbie Hallig

Mary W. Harrison

Greenbrier Farm and Gardens
Carl Phelps and Nan Jenks-Jay

Lenox Garden Club

Elizabeth Neyland

Julie Richburg

Emily M. Rose

William and Martha Selke

Sheffield Land Trust

Stockbridge Land Trust

Sweetwater Trust

The Berkshire Garden Club

The Nature Conservancy-Berkshire Office

The Trustees of Reservations

Williamstown Rural Land Foundation

Windy Hill Farm

TABLE OF CONTENTS

FOREWORD

A modern flora of one of the state's most interesting and important botanical regions is a publication that many botanists have been eagerly awaiting. In *Flora of Berkshire County, Massachusetts*, Pamela Weatherbee has meticulously reported all of the plant species known to occur in the range of habitats that the county's geologic and topographic diversity provide. As it offers an accurate record of the current status and distribution of the county's plant species, this flora also provides valuable insights into floristic changes that have occurred since Ralph Hoffmann's *Flora of Berkshire County* was published in 1922.

The Natural Heritage & Endangered Species Program, which conducts biological surveys of plant and animal species in the Commonwealth, particularly for rare species, has since 1982 benefited greatly from Pamela Weatherbee's botanical exploration of Berkshire county. She is without a doubt one of the state's best field botanists, quietly and painstakingly documenting weed species as well as rarities. A member of the New England Botanical Club since 1985, she has also been very active with a variety of conservation groups ranging from those acquiring land to those concerned about the impact of non-indigenous plants to one dedicated to protecting Mount Greylock.

Berkshire County now joins less than a dozen of New England's sixty-seven counties for which a comprehensive flora has been prepared over the last thirty-five years. For the benefit of present and future scientists, conservationists and naturalists, as well botanically inclined visitors to the Berkshires, Pam Weatherbee's *Flora* makes an enormous contribution to the knowledge of this diverse and fascinating area.

Henry L. Woolsey
Natural Heritage & Endangered Species Program
Massachusetts Division of Fisheries & Wildlife

Introduction

Berkshire County, Massachusetts, is a region of considerable botanical diversity, due in part to varied topography and to bedrock geology that includes the only extensive marble deposits in the state. Data from the Massachusetts Natural Heritage and Endangered Species Program (MNHESP) indicates that the county has more state-listed rare species than any other county in the state. This is a good indicator of a variety of intact habitats, which are the product of a complex of factors discussed in this flora.

Ralph Hoffmann produced the most recent flora in 1922, and he had only Dewey's flora of 1829 to work with. Hoffmann's flora was an excellent complete list, with concise descriptions of habitats and distribution of plant species. Much has changed in the last 72 years. A renewal of interest in botany and the preservation of rare species has resulted in the discovery of many new native and alien species in the county. Possibly some species have disappeared from the county, although this is harder to document. There has been an expansion of knowledge of the origins of the flora and events after glaciation. The county also has undergone a change in land use, seen both in increased development and in changing patterns of agriculture.

With this in mind, it became obvious that a new flora of the county was needed; to modernize the taxonomy and classification, to record new species and search for others previously recorded, and to investigate and explain, at least superficially, the causes of the diversity of the flora. It is fairly easy to record what plant species are here but impossible to understand completely why those species are here and why they grow where they do, given the difficulty of discovering past events. This account can be only a brief summary of the findings of the many scientists who have gradually unfolded the origins and migrations of our flora. This most recent writer, who sometimes has literally walked in their footsteps, must rely heavily on all those who have gone before.

To produce this updated flora the author conducted intensive field work from 1986-1994, although some collecting and study began in 1982. Major habitats, and as much of the county as possible, were explored and information recorded on vegetation and plant communities. Specimens of most species were collected, and have been deposited in the Hodgdon Herbarium of the University of New Hampshire and in the author's personal herbarium. However, not every inch of the county has been scoured, and not every species has been rediscovered, which leaves that pleasure and excitement to the future.

All species, except for two or three, are documented by specimens: a majority of them collected recently, some before 1970. Various herbaria, mainly the New England Botanical Club (NEBC), Gray Herbarium (GH) of Harvard University, Hodgdon Herbarium of the University of New Hampshire (NHA) and North Adams State College (NASC), were searched for specimens and locations. As noted in the acknowledgments, many other people were of invaluable help in this work.

A flora can only be a comprehensive picture of the vegetation seen at the time and represents all that is known up to that point. As soon as the last word is printed, the picture will have changed. New species may have migrated into the county, while others may have dropped out. Thorough searching or chance encounter may turn up plants that were here all along. In response to climatic change, whole plant communities may be shifting their range. While change may occur over a time span not easily seen during our own lifetime, we can be sure of that one event: change. A flora is a work in progress.

A county flora covers a small area, but it is a basic building block in the assembling of information on the vegetation of a larger region. It is an area small enough to be studied closely, but large enough to include many of the plant communities seen in the state and a good percentage of plant species seen in the northeast. Political boundaries do not often follow limits of natural conditions, but at least provide a frame. And, in actuality, the geology and topography of this county do agree with the political boundaries in a general way.

The photographs were selected with an emphasis on the State-listed rare and more uncommon species. They were taken by the author except where noted. All photographs were taken within Berkshire County.

Berkshire County, located along the western boundary of Massachusetts, is approximately 53 miles (85.3 km) long and an average of 17.5 miles (28.2 km) wide and covers approximately 940 square miles (USDA, 1988). It is bounded on the north by Vermont, on the west by New York and on the south by Connecticut.

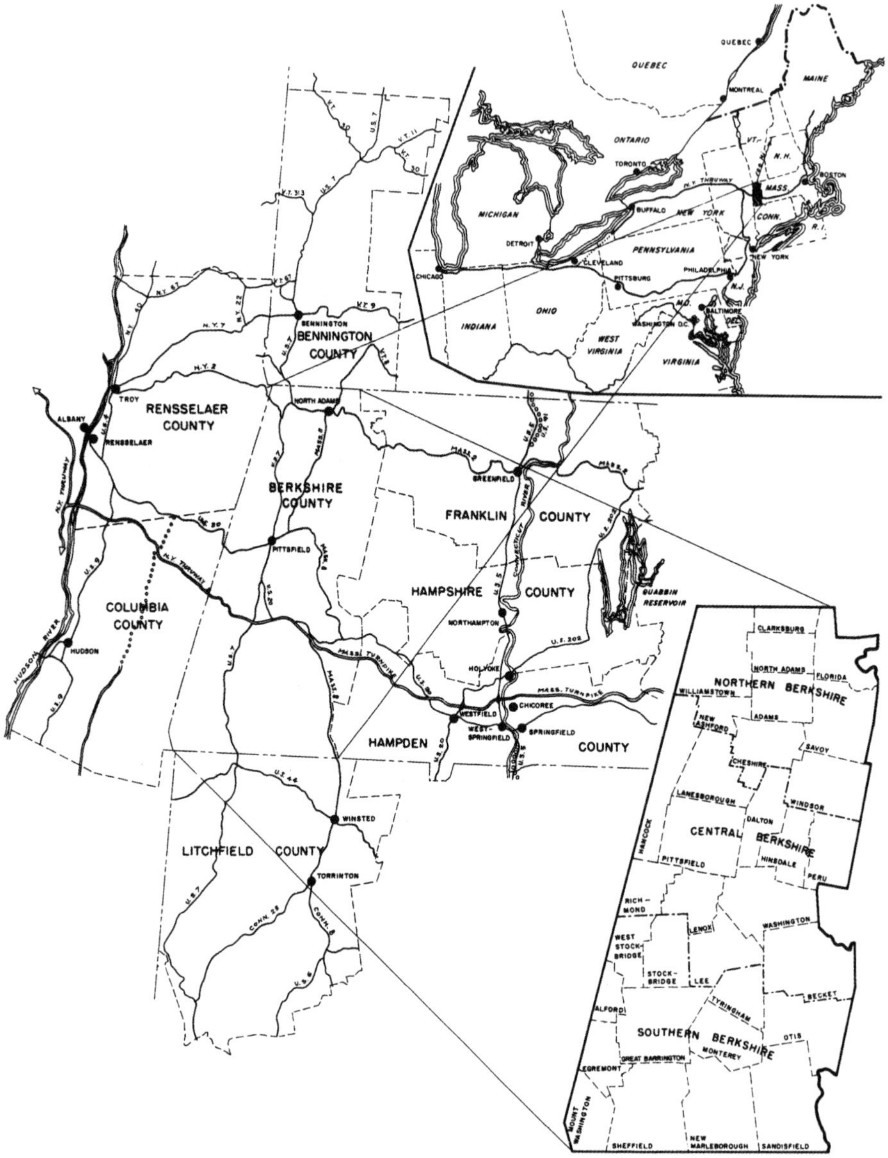

Figure 1. Location map for Berkshire County
(From Technical Planning Associates, 1959)

FLORA OF BERKSHIRE COUNTY
MASSACHUSETTS

I. Factors that Influence Vegetation

1. Topography

Topographical diversity is an important factor in the formation of many microhabitats that influence the growth and survival of the vegetation of an area. The type of topography affects the amount of soil accumulated in an area, and the water, sunlight and heat received by the vegetation. High elevations are colder and wetter than the valleys. Vegetational gradient analyses (Siccama, 1974) have shown changes, usually gradual, in relative abundance and diversity of species over change of elevation. East-west trending ridges provide north-facing and south-facing slopes with greatly differing conditions and, therefore, different plant communities, growing in close proximity (McVaugh, 1958). Wide valleys with meandering rivers allow alluvial deposits to build up in floodplains. In sharp contrast are narrow ravines and valleys with fast-flowing rivers and little floodplain. Topographical variation provides the basins and low areas for swamp and marsh development.

The topography of Berkshire County includes mountainous terrain, rolling plateaus, and both narrow and broad valleys. Thirty-two percent of the county has slopes steeper than 15% (Technical Planning Associates, 1959). The overall pattern of topographical features can be summarized as a system of central valleys bounded on the east and west by north-south trending mountain ranges. Four main physiographic regions, intimately related to the bedrock geology, comprise the county.

The high Berkshire Plateau, stretching from Vermont to Connecticut, forms the eastern half of the county. The Taconic Range extends along the western border between New York and Massachusetts, with numerous ridges running into the Central Valley region, which occurs between the two mountainous areas. The Green Mountain Region is a small southerly extension of the Green Mountains of Vermont into the extreme northern part of the county.

The greatest extremes of elevation are found in the northern third of the county, in the towns of Williamstown, Clarksburg, North Adams, Adams, Cheshire and New Ashford. Mount Greylock, at 3491′ (1064 m) the highest point in the state, and associated mountain ridges averaging 3000′ are bounded on three sides by low valleys. Eight miles away, the north-flowing Hoosic River flows into Vermont at elevation 577′ (176m). The valleys of this section are narrow and steep-sided with small areas of level floodplain and marsh.

The Taconic Range, which extends southward along the western border, consists of a line of summits ranging from 2168-2795′ (661-852m). Its eastern slopes are dissected by many narrow ravines. An isolated outlier of the Taconics, the Greylock Range has two ridges that enclose on three sides a steep-sided valley known as "The Hopper" from its resemblance to a grain hopper as seen from the west.

An abrupt escarpment known as the the Hoosac Range forms the western border of the Berkshire Plateau, a well-defined area that is topographically and geologically distinct from the rest of the county, as was noted by Egler (1940) and Hoffmann (1922), who referred to it as the Hoosac Plateau. The highest points of this plateau, along the western edge are about 2700′ (823m) but the average elevation is 2000′ (610m) gradually becoming lower toward the Connecticut border. The plateau has a rolling indeterminate topography, cut by narrow ravines of the tributaries of the Deerfield River.

From Lanesboro south, the variations of elevation become less extreme, the valleys wider and slopes less steep. The major central valley of the southern two-thirds of the county are associated with the Housatonic River, which rises in Lanesboro. Because of gentle slopes and abundant water power, this valley became the most populated and developed both industrially and agriculturally. The valley is broken up by numerous hills and ridges which create many smaller valleys to the east and west. In Sheffield this wide valley is bounded on the west by the sharp escarpment and summits of Mount Everett at 2598′ (792m) and Mount Race at 2362′ (720m). The Housatonic River flows out of the county at elevation 656′ (200m). Williamstown and Sheffield share the distinction of the lowest elevations of the county.

2. Watersheds, Water Bodies and Wetlands

Seven river systems drain three watersheds (Figure 2). The Hoosic River and Bash Bish and Kinderhook Brooks empty into the Hudson River. The Deerfield and Westfield Rivers flow into the Connecticut River, while the Farmington River eventually joins with the Housatonic River flowing into Long Island Sound. Except for the Deerfield, these river systems originate mainly in the county.

Rivers and streams offer continous migratory routes for plants through the banks, wetlands, floodplains and low valleys that they have created. They also provide a variety of habitats. Flowing as they do into widely separate regions, Berkshire rivers provide migratory routes from these areas, which may add to the diversity of species seen here.

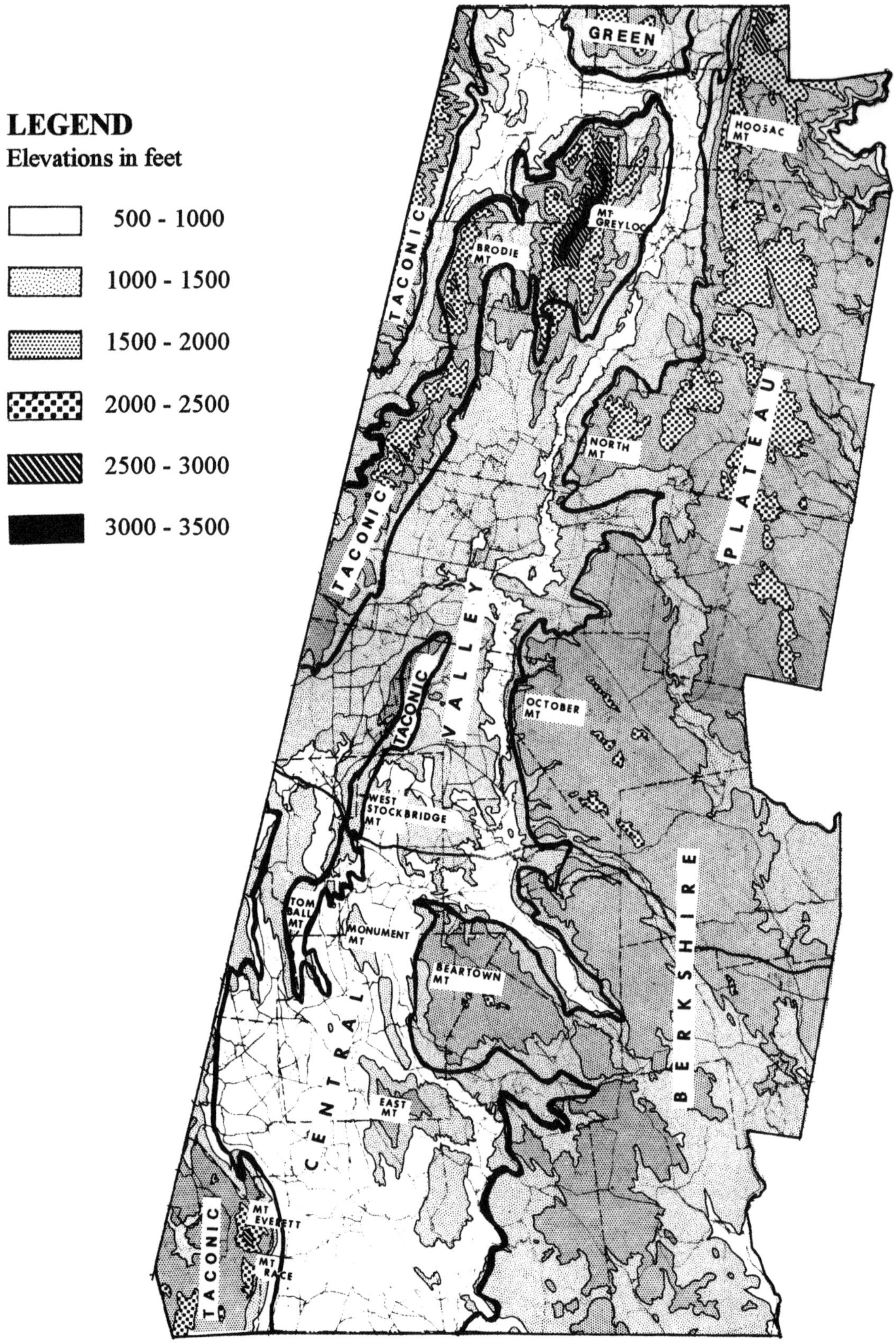

Figure 2. Topography and physiographic regions of Berkshire County (adapted from Technical Planning Associates, 1959)

There are approximately 197 lakes and ponds in the county (McCann and Daly, undated). Forty-five percent of these are artificial ponds or reservoirs, and many others have been enlarged, enhanced or altered in some way, either to supply water power to mills, or for recreational purposes.

Because of its rolling topography, the Berkshire Plateau has many ponds, bogs and marshes. North and South Ponds in Savoy are natural ponds, while Bog and Burnett Ponds are examples of bogs or boggy areas dammed for recreational purposes by the Civilian Conservation Corps (CCC) in the 1930s. The town of Washington has several lakes created for water reservoirs, and also abundant marshes and bogs. Becket and Otis have many recreational lakes and extensive natural marshes and bogs. Big Pond in Otis is a good example of a natural large pond with a sandy subtrate.

The Taconics have a few ponds, notably Berry Pond, reputedly the highest natural body of water in Massachusetts at 2080′ (634m). Guilder and Plantain Ponds, at the base of Mount Everett and Mount Race are enhanced ponds.

The northern towns of the central valley have few natural ponds and no lakes, but several reservoirs and impoundments, such as Cheshire Lake. Starting at Lanesboro, major bodies of water such as Pontoosuc and Onota Lakes, Richmond Pond, Stockbridge Bowl, and Laurel Lake swing south and drain into the Housatonic River. Lakes Garfield and Buel, further east, are in the Konkapot River watershed.

Three percent of the county is wetland (Technical Planning Associates, 1959). Seven northern valley towns have an average of 0.9% of their area in wetlands, while Stockbridge and Sheffield are 12% wetland. Low areas along the Housatonic River create most of these marshes and swamps, but there are also shallow basin marshes, headwater swamps, fens and marshes associated with lake margins.

3. Geology

Bedrock geology is a major factor in determining the vegetation and species diversity of a region, as are the climate and the pool of species available to colonize the habitats. The forms of the land, the slopes, valleys, mountain summits, drainage patterns and the habitats that develop in these diverse areas, are the result of geologic and climatic forces acting upon the various types of rock.

The soils that develop from different types of bedrock often determine what species and plant communities will occupy those sites. Springs and other ground water coming to the surface are influenced by rock formations and their mineral composition, which affect the acidity or alkalinity of the water. This, in turn, has an important effect on what types of communities will occur where these waters surface.

The geology of the region is complex, and only recently have the origins of the formations been completely worked out. Western Vermont and northwestern Connecticut share a similar history. Most of the following description is based on work by Ratcliffe (1979), Stanley and Ratcliffe (1985) and Van Diver (1987). The latter contains a simplified and condensed account of some of the work of the former two authors as it applies to Vermont, which would also apply to Berkshire County.

The oldest formation is the Berkshire Massif, gneisses that are the remnants of eroded mountains approximately 1.4 billion years old and were the basement rocks of an ancient continent. These rocks make up most of the Berkshire Plateau and appear in the Green Mountains of Vermont and the Adirondacks of New York.

About 650 million years before present (BP), rifting began to produce a proto-atlantic ocean and by 520 million years BP warm shallow seas covered most of what is now western New England and New York, and bordered the ancient continental rocks. Sandstones formed from beach deposits, limestone from reef organisms, and shales from mud deposited in deeper water, all of which built up over the continental shelf. This tranquil time of continued deposition persisted until about 470 million years BP, when during Ordovician time the proto-atlantic ocean began to close.

This was the beginning of the Taconian Orogeny (mountain-building), when pressure from the oncoming African Plate created volcanic island arcs offshore. When the two plates collided, the island arcs and oceanic crusts were crushed against the shore. Serpentine deposits were formed from oceanic crust fragments that were moved along by seafloor spreading and crushed into or over the continental margin (Stanley & Ratcliffe, 1985; Dann, 1988). Continued pressure forced the folding and faulting of rock formations, which took place under miles of rock. Heat and pressure metamorphosed the rock; sandstones became quartzite, limestone became various kinds of marble, and shales metamorphosed into schists. Intense pressure from the east thrust slices of rock westward 20-30 miles, and stacked them so that they rested on top of the marble and quartzite. These schists now form the upper elevations of the Taconic Range, Mount Greylock Range, Mount Everett and Mount Race, Tom Ball and West Stockbridge Mountains and June Mountain. They cap small hills and ridges, like Bear's Den and Bartholomew's Cobble in Sheffield.

During later mountain-building about 390 million years BP, the Precambrian continental shelf basement rocks were raised up and are now exposed as the Berkshire Massif, a major part of the Berkshire Plateau. The conjoining of the Eurasian, North American and African continents formed Pangaea, a single landmass, which began to break up around 200 million years BP. About 140 million years BP flowering plants (the Angiosperms) arose, becoming the dominant vegetation by 60 million years BP. During that time land bridges that still connected Eurasia and North America through Greenland and between Alaska and Siberia allowed plant migration which made possible the widespread similarity

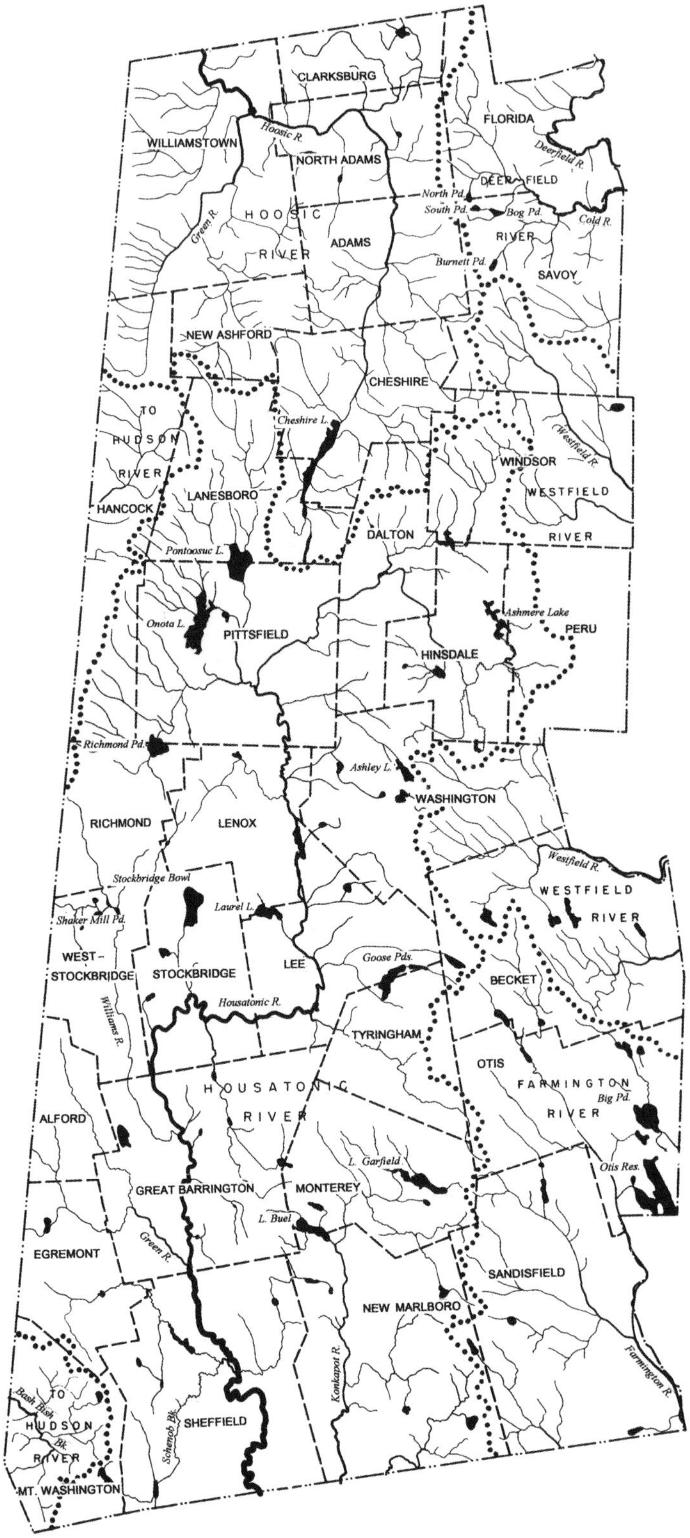

Figure 3. Watersheds and Drainage of Berkshire County (adapted from Technical Planning Associates, 1959).

of plant species composition (Graham, 1993). The southern portions were splitting away, and by 20 million years BP the land bridges in the northern sections of these continents were broken.

Erosion of miles of rock created the major land forms as they are seen today. Resistant rocks, either quartzite, schist or gneiss, cap the summits of mountains. The soft easily-eroded carbonaceous rocks (marbles) are found on lower slopes and valley floors.

It is not known what the pre-glacial topography was like 2 million years BP, but it was probably more rugged, with sharp peaks and deep V-shaped erosional valleys. Glacial action modified but did not change the basic mountain-valley formations. The ice sheet, moving from NNW to SSE, rounded mountain tops and created gentle U-shaped valleys, some of which became partially filled with glacial debris. Up to 18 incursions of ice (Davis, 1983) pushed down from the north, but the impact of only the most recent, the Wisconsinan, can be seen today.

As the ice sheet pushed along, it carried a burden of boulders, gravel and clay, which was deposited as a blanket when the glacier melted. The ice rode up the northern ends of ridges, but plucked rocks from the southern ends, thus creating south-facing cliffs which can be observed at Jones' Nose and Rounds Rock on Mount Greylock, as an example. Huge boulders were deposited as "Balance Rocks" as in Lanesboro and Savoy. Glacial erratics were scattered over the landscape. Boulder trains were deposited in Richmond, Sheffield and New Marlboro.

As the glacier melted back, around 18,000 BP, the mountain summits were uncovered first. Tongues of ice remained in the valleys (Bierman & Dethier, 1986) where they blocked runoff, forming small lakes where sand and gravel accumulated. Terraces were formed from the accumulation of till at the edges of the ice along the sides of the valley.

In the northern Hoosic Valley, glacial Lake Bascom was formed by blockage of the north-flowing Hoosic River by the retreating ice margin, which stagnated approximately 12 miles (20 km) downstream. Its highest level was 1040′ (317m) and filled the valleys of Williamstown, North Adams and Cheshire. The hills of glacial till were not stabilized by vegetation until around 12,000 BP when tundra vegetation and scattered jack/red pines became established (Whitehead, 1979). Erosion was extensive, and layers of clay and silt accumulated rapidly on the lake bottom.

A glacial lake, Lake Ashley, existed along the Housatonic River valley from Falls Village, Connecticut to Great Barrington, with lacustrine deposits up to 240′ (73m) (Holmes & Newman, 1971). Sandy outwash borders the river in Sheffield. A wide valley and gentle grade has allowed the river to build up thick layers of alluvial soil in the floodplains.

In summary, geologic forces have caused several types of bedrock to become intertwined and compressed into this small area, Berkshire County, creating the possibility of many plant communities coexisting close together. No one community is very large, however, as compared to an extreme example, the prairie.

Glaciation removed all prior vegetation and reworked the soil, therefore our flora is younger and poorer in species than that of the southern Appalachians. However, glacial processes have left a number of formations and soil types that favor the growth of different plant communities. Sandy outwash, not a common formation in the county, is colonized by pitch pines (*Pinus rigida*) and other sandplain species. Deltas, created by streams or deposits of gravel, support species that demand well-drained situations. Underlying clay deposits create poorly-drained areas that become marshes or bogs. Glacial action carved out basins that were conducive to the formation of bogs or fens. Some mountain tops received a layer of till, while others appear to have been scraped bare.

4. Climate and Weather

Climate is a major controlling force in determining the vegetation of a region (Woodward, 1987). The temperature, particularly the minimum, is a controlling force in the type and species of vegetation seen. Broad-leaved deciduous forest is expected where the minimum does not fall below -40° F.The average annual rainfall and the timing of high and low precipitation are important factors. For example, if most of the rainfall occurred during the cold months, as rain or snow, little moisture would be available during the growing season, and our vegetation would consist of dry scrub or desert plants. Closed canopy forest at our latitude requires at least 30 inches (76 cm) of rainfall evenly distributed.The number of frost-free days (growing season) is also critical in limiting plant distribution (Siccama, 1974). Vegetation is also affected by the frequency of catastrophic events such as droughts, floods, windstorms, and ice storms. The overall climate of Berkshire County is composed of a mosaic of microclimates which are products of local elevation and topography. Due to lack of long term data for many locations, it is difficult to document these differences.

The climate of Berkshire County is similar to the rest of inland New England in that it is largely continental, with cold winters and hot summers (USDA, 1988). Supply of moisture usually exceeds evaporation except during severe droughts, which are fairly uncommon.

Data from Stockbridge for the period 1951-1974 (USDA, 1988) give an overall view of the climate of the valley region. Table 1 summarizes temperature data. The average total rainfall is 43 inches (109.2 cm) but may range from 30 inches (76cm) to 68 inches (172 cm) and is fairly evenly distributed throughout the year, with 55% falling from April through September. Snowfall averages 71 inches (181.4 cm) typically falling from December to March. The frost-free (growing season) ranges from 103-144 days. The term

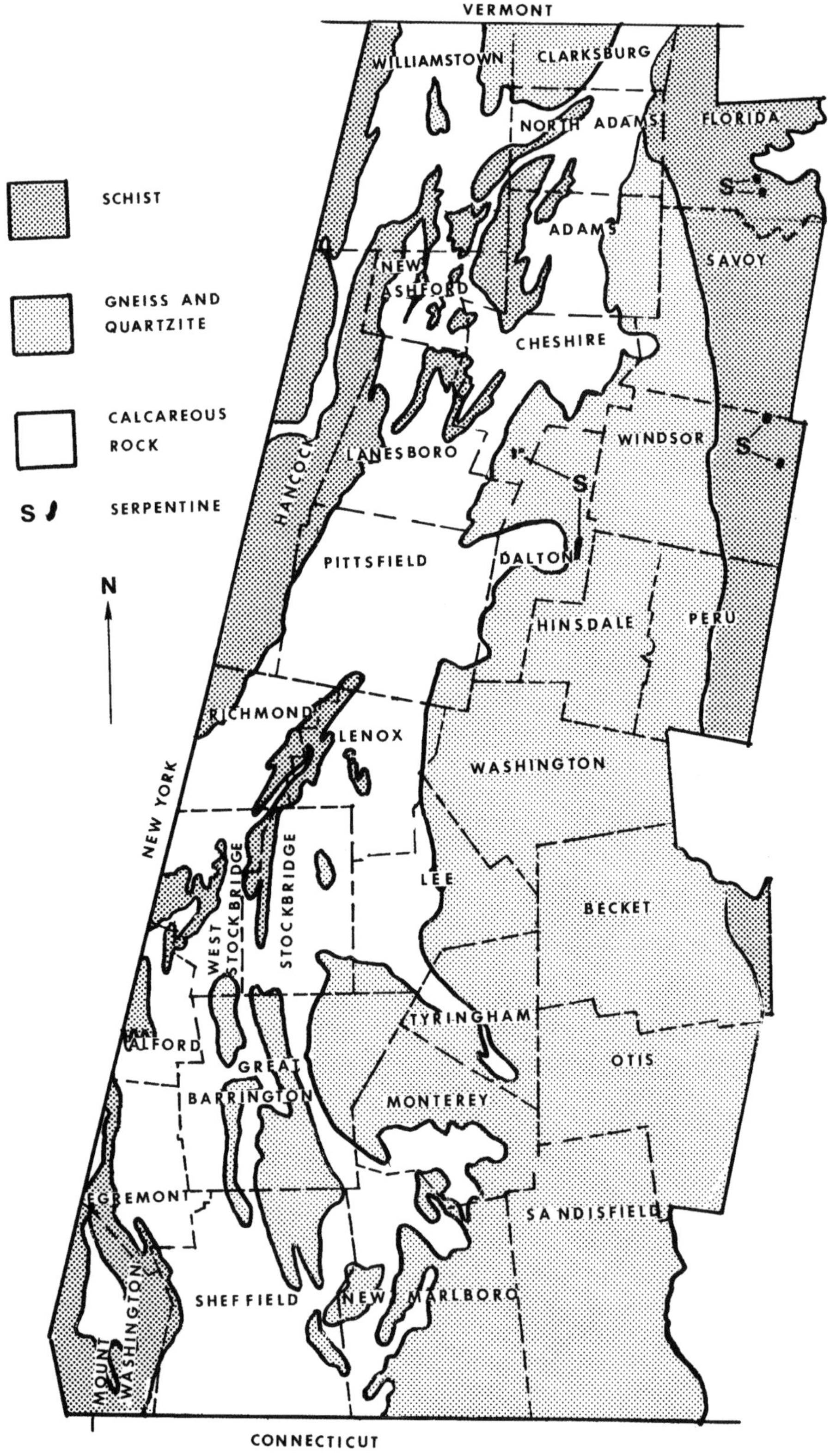

Figure 4. Bedrock geologic map of Berkshire County
(redrawn from Zen, 1983).

"growing season" is more applicable to cultivated crops rather than to natural vegetation. Growth begins in March when willow catkins emerge and skunk cabbage (*Symplocarpus foetidus*), producing its own heat, melts its way through frozen marshes. First frosts do not discourage asters and fringed gentian (*Gentianopsis crinita*) from flourishing until late October. Native evergreens will take advantage of warm early spring or late fall days to photosynthesize (Marchand, 1987).

Table 2 summarizes climatic data available for five towns. Perhaps the most significant figures are the length of the frost-free season and average rainfall, which correlate closely with elevation.

Records have been kept at Williamstown virtually continuously from 1816 to 1951 (Milham, undated), and indicate that the average yearly temperature was 45.7° F (7.6° C) and the average annual precipitation was 38 inches (96.4 cm). Averaged over a long period, temperature and rainfall vary little, but as every New Englander knows, from day to day, month to month or year to year there are wide, not to say wild, variations.

Records have been kept at the Pittsfield Airport for 54 years and were kept at North Adams from 1900 to 1979. Recent highs and lows gleaned from these records show how trends cannot be seen from a look at a short time span. The summer of 1988 was one of the most stiflingly hot summers, which led to speculation on whether global warming was setting in. December 1989, with an average of 11.7° F (-11.3° C), was the coldest in 50 years, but was followed by the warmest January on record averaging 30.9° F (-1.1° C). The year 1990 went on to prove the wettest ever with 68 inches (172 cm) falling.The summer of 1992 was cool, with a record cold July averaging 65.5° F (18.6° C). January 1994 averaged 12° F (-11.1° C) and January 27 broke the record for coldest with -26° F (-32° C), while unofficial readings around the county hit -30° F. Within 24 hours the mercury had risen to 56° F. Over thousands of years, the vegetation has either adapted to such variations or dropped out.

Catastrophic events, such as floods, windstorms, ice storms and drought, are infrequent, but as disturbances that affect vegetation they play an important part in the creation of many different habitats for diverse plant communities. The overall climate produces conditions that result in closed canopy broadleaf deciduous and needleleaf evergreen temperate forest but catastrophic events cause heterogeneous patches with diverse vegetation.

Flooding, when uncontrolled, deposits new layers of soil on floodplains, banks and sandbars, continually creating open areas for colonization by plants. It encourages certain species while eliminating others. Ice scouring of shores creates riparian meadows, which were noted by early settlers along the Housatonic River. With increasing control of the river's natural processes comes a decline in these habitats.

Table 1. Temperature data for Stockbridge for the period 1951-1974 in degrees Fahrenheit.

	Average daily max.	Average daily min.	Average daily	Highest for period	Lowest for period
January	31.8	11.9	5.9		-29
July	80.1	56	68.1	95	
Annual			45.7		

From U. S. D. A. (1988).

Table 2. Climatic data for five towns. Temperature in degrees Fahrenheit. Precipitation in inches.

	Williams-town	Lanesboro	Great Barrington	Savoy	North Adams
January average daily temp.	23.2	17.4	22.8		*26.1
July average daily temp.	69.6	66.6	68.7		*71.1
Average annual temp.	45.7	43.7	45.7	44.9	*46.6
Growing season days	113-140	111-133	108-151	<120	
Average annual precip.	37.9	45.6	44.8	49	43.8
Elevation in feet	708	1,198	699	1,886	1,198

Sources of climatic data

Williamstown: Milham (undated), data from 1817-1951.
Lanesboro: U. S. Department of Commerce, data from 1981-1987.
Great Barrington: U. S. Department of Commerce, data from 1981-1987.
North Adams: Precipitation data taken at the reservoir (El. 1198′). From Annual Reports, City of North Adams, City Engineer's Report, 1900-1979. *Temperature data taken at the airport (El. 652′). From New England Weather Assoc., 1920-1979.
Savoy: From Department of Environmental Management (undated, unpublished material for GOALS Plan for Savoy State Forest).

Heavy rains that accompany flooding, or an unusually wet season such as in 1990, loosen soil that often slides off steep slopes. Two major landslides occurred on the slopes of Mount Greylock in 1990, leaving open soil which will be colonized by pioneer species such as paper birch (*Betula papyrifera*), pin cherry (*Prunus pensylvanica*) and red maple (*Acer rubrum*). Former landslide sites on the slopes are noticeably different in vegetation.

Windstorms may be wide-ranging in their effect or concentrated in one locality. Hurricanes have struck occasionally, most notably in 1938 and 1955, but did not cause total destruction of forests, as the 1938 storm did in the rest of New England. More frequent, violent tornado-like winds that affect a small area, even a few acres, are important in opening small patches to sun-loving pioneer tree species and herbaceous plants. A tornado-like wind destroyed an old growth stand of white pine (*Pinus strobus*) and eastern hemlock (*Tsuga canadensis*) in adjacent northwestern Connecticut (The Nature Conservancy News, 1989).

Ice storms, as well as early or late snow storms, can damage vegetation, opening the canopy of a forest. An early snowstorm on October 4, 1987 left 15 inches (40 cm) of heavy snow on trees still in leaf. Branches and crowns of large trees were broken, but smaller trees were crushed and uprooted. Damage was minimal above 1500′ (457m) where the snow was powdery. This was rated as a minor disturbance since it did not change the dominant canopy trees (Potter, 1989).

Rainfall may vary 20-30 inches (50-76 cm) from year to year but severe drought is rare. In a dry period from 1961-1966, yearly totals of precipitation averaged 11 inches (29.7 cm) below normal (USDC, 1960-1966), and most importantly, lower monthly rainfall occurred in the growing season when evapo-transpiration is highest. Even one dry season may retard usual succession to closed canopy in a normally xeric (dry) rocky habitat.

If any one of these catastrophes became more frequent, it would affect the types of vegetation normally growing in this region. The climate undoubtedly is changing in a directional way over a long period of time (Davis, 1986), but there would be a time lag before any change in vegetation would be seen. Our plants exist in a delicate balance of many complicated, interacting forces.

5. Origins and Floristic Affinities of the Flora

Given the physical factors affecting the vegetation, discussed above, the other important influence on the families and genera of plants found today in Berkshire County is the pool of species that was available to colonize the area. The development of the Tertiary flora and subsequent glaciation has resulted in the species seen at present in the floristic province of which Berkshire County is a part.

Floristic Province.

Berkshire county lies within the Appalachian Province of the North Atlantic Region as defined by Takhtajan (1986) and by Cronquist (1982). This corresponds generally to the area described by Braun (1950) as the Eastern Deciduous Forest and to the Eastern Deciduous Forest Province of Gleason and Cronquist (1964). The most widespread and characteristic type of vegetation is the deciduous forest which includes, among others, these found in Berkshire County: sugar maple (*Acer saccharum*), red maple (*A. rubrum*), American beech (*Fagus grandifolia*), red oak (*Quercus rubra*), white oak (*Q. alba*), paper birch (*Betula papyrifera*), black birch (*B. lenta*), yellow birch (*B. alleghaniensis*), shagbark hickory (*Carya ovata*), American basswood (*Tilia americana*), white ash (*Fraxinus americana*), black cherry (*Prunus serotina*), butternut (*Juglans cinerea*), cottonwood (*Populus deltoides*), big-tooth aspen (*P. grandidentata*), elm (*Ulmus americana*), and sycamore (*Platanus occidentalis*). Important evergreen components of this region are the conifers, white pine and eastern hemlock (*Tsuga canadensis*). An abundant understory layer, an herbaceous ground cover and, in favorable soils, a rich vernal flora, are integral parts of the deciduous forest.

Since Berkshire County is situated in the northeastern section of the Appalachian Province, where it is relatively close to the both the Atlantic and Gulf Coastal Plain Province and the Canadian Province (Takhtajan, 1986), it is not surprising that plants typical of these provinces make their appearance in the county. Mountainous regions, up to 3491′ (1064 m), provide climatic conditions similar to the Canadian Province. Sandy outwash plains and acidic bogs and marshes are habitats suitable for the few Coastal Plain species.

One of the most interesting aspects of the Appalchian Province is the great number of taxa that are very similar to those of the Eastern Asian Region. The following genera, which are found in Berkshire County, occur only in eastern United States and in eastern Asia (Li, 1952): *Camptosorus, Carya, Caulophyllum, Hydrastis, Liriodendron, Penthorum, Hamamelis, Parthenocissus, Panax, Aralia,* and *Epigaea*. Most of these genera, except for *Carya*, consist of one or two species in eastern United States, and several to many in eastern Asia. The physiognomy of both regions is deciduous forest.

Glaciation and its Effect on Vegetation.

The greatest single event that has determined the composition of our flora is glaciation. Glacial periods eliminated or dispersed the widespread, more homogeneous, deciduous and sub-tropical Tertiary flora of the Northern Hemisphere, resulting in a mixing of both northern and southern elements (Wood, 1970).

During the early and middle Tertiary period, at least 50 million years BP, a broad band of vegetation known as the Arcto-Tertiary Flora covered the Circumpolar region (Chaney, 1947). This forest contained many genera of our present deciduous forest such as *Fagus* and *Quercus* (Graham, 1993 in *Flora of North America*) and some genera now found only in Eastern Asia, such as *Metasequoia* and *Cercidiphyllum*. An exchange of species between Asia, North America and Europe was possible at times over the Bering land bridge and the North Atlantic land bridge. The climate was warm temperate, although North America was approximately in the same position as at present.

The Brandon Lignite, found in North Central Vermont, is a Tertiary deposit dating from 32-34 million years BP which has yielded macrofossils that indicate a flora most similar to modern floras of southeastern United States and southeastern Asia.

A cooling trend in the late Tertiary, 27-2 million years BP, resulted in a temperate deciduous forest that supplanted the warm-temperate forest at higher altitudes (Larsen, 1980), and which was rapidly replaced by Arctic and Boreal species now found in the north. This cooling trend culminated in the formation of persistent ice in the Polar regions.

The Quaternary Period, at the beginning of the Pleistocene Age at 2 million years BP, saw the extensions of these glaciers southward over portions of the landmasses of North America, Europe and Asia. Up to 18-20 glacial advances of approximately 100,000 years' duration (Davis, 1983), interspersed with short interglacial warm periods of 10,000-15,000 years duration, characterized the period. Vegetation repeatedly migrated southward during the long cold periods, and then migrating north during the short warm intervals. Climatic changes were fairly abrupt (Davis, 1983), which resulted in much vegetational instability. Recent work, inferred from pollen records, has established that plant species migrated individually, not as plant communities. Plant communities that formed south of the glacial margin were made up of assemblages not seen today (Pielou, 1991).

Return of Vegetation after the Most Recent Glaciation.

Using pollen profiles from cores taken from bogs and ponds throughout eastern North America, Davis (1983) has traced the return of some of our important trees from their refugia in the south or on the coastal plain. Determining accurately the vegetational composition of an area from pollen deposited in ponds and bogs is difficult, as some species release large amounts of pollen, like oak, birch and pine, while maple, fir, poplar, larch and basswood do not (Davis & Goodlett, 1960).

According to Davis, spruce (*Picea* spp.) migrated north from a widespread distribution in the south, into the Appalachian Mountains and northeastward into New England. Larch (*Larix laricina*) persisted only on the eastern edge of the Great Plains, and moved northeast from west of the Appalachians. The jack/red pine (*Pinus banksiana/P. resinosa*) group migrated the coastal plain and Appalachians northward as did fir (*Abies balsamea*). These conifers, arriving in New England c. 12,000 years BP, were among the earliest trees to colonize the recently deglaciated region, and now have northern distributions. White pine and eastern hemlock migrated from the coastal plain, arriving in New England 10,000-9,000 years BP.

Deciduous trees, such as oak, elm, maple, hickory and American chestnut (*Castanea dentata*), migrated from their refugia in south central United States, arriving at different times. Oak reached New England c. 10,000 years BP while American chestnut arrived only 2000 years BP and became a major important tree in eastern forests, only to be decimated by chestnut blight in the early 1900s.

Berkshire County was totally glaciated, the ice limit of the Wisconsin glacier being some 150 miles to the south. The maximum occurred at 20,000 BP. Climatic changes caused rapid melting, and the county was ice-free by 14,000 years BP (Davis and Jacobson, 1985).

Whitehead, a professor at Williams College in the 1960s, made a detailed study of the pollen deposits in cores from Berry Pond, Pittsfield State Forest (Whitehead, 1979). Berry Pond is a small natural pond in the Taconic Range at elevation 2080′ (634 m). He found that the sedimentary record began about 14,000 years BP and parallels closely pollen profiles seen at other New England sites.

His findings are similar to those of Davis. The first type of vegetation was probably tundra with sedges dominant and with scattered jack or red pine. From about 12,000 to 10,200 years BP, spruce became common, and alder (*Alnus*), juniper (*Juniperus*), larch, fir, oak, and hornbeam (*Ostrya/ Carpinus*) were well represented. Presence of spruce has been corroborated by the discovery of macrofossils of white spruce cones (*Picea glauca*) in a fen in Egremont, which were carbon-dated to c. 11,630 years BP. Conifers were the preferred food of mastodon, bones of which were found in the same fen (Parrish *et al.*, 1984; Moeller, 1984).

After 10,200 years BP, pine and birch became abundant, with white pine common, but some pitch pine (*Pinus rigida*) and eastern hemlock were also present. Oak and other deciduous trees such as sugar maple, white ash, and elm increased in abundance. From 8,250 years BP to the present, oak, birch, beech, and eastern hemlock characterized the area. American chestnut arrived later, and became common around the pond. Whitehead's pollen profile shows a dramatic decline of hemlock around 4800 years BP, which corresponds to a similar decline noted all over the northeast. A disease or insect outbreak may have caused the decline. At present, red maple, red oak, and beech are dominant around the pond, with sugar maple, yellow birch, black cherry, paper birch and white ash common also. Few shrubs and herbaceous plants are represented in the pollen record, and none in recent sediments, so we must infer that their arrivals occurred at different times, as did the trees.

Floristic Affinities.

Repeated displacement of species during the Pleistocene, their immigration back into regions of bare soil, and the subsequent mixing of species of quite different regions, from the boreal to the southern and of different vegetational associations, has resulted in a present-day flora that may appear to be a confusing mixture. Davis (1983) calls the Eastern Deciduous Forest a "grab-bag of species" which was selected for its colonization and competitive abilities. It is not surprising that our forests are able to renew themselves after repeated timbering and clearing for agriculture.

Each species has an individual distribution, but patterns emerge indicating that many species share a more or less similar distribution with a geographic center which determines its name or floristic affinity. The present area of each element strongly suggests that it is related to the location of refugia of species of that element during glaciation, and conditions soon after glacial retreat.

The following floristic groups are found in Berkshire County: (1) Eastern Deciduous Forest element (including a southern and mid-western element), (2) Alleghanian element, (3) Boreal element (including circumpolar plants), (4) Coastal Plain element.

A range map of a species with a typical distribution of that floristic group illustrates the area of each element. Common species typical of each element will be discussed and distribution of selected rare species will be used to illustrate the affinities and migration patterns of Berkshire County flora. Possible causes of rarity are considered. Rare taxa are defined as only those species listed by the Massachusetts Natural Heritage and Endangered Species Program (Natural Heritage, 1994).

Eastern Deciduous Forest Element.

Since Berkshire County lies within Braun's (1950) boundary of the Eastern Deciduous Forest, obviously plants of this element are among the most common and important to the general character of the vegetation. They include most of our deciduous trees such as sugar maple (Figure 5), red maple, silver maple (*Acer saccharinum*), red oak, white oak, black cherry, American basswood, tulip tree (*Liriodendron tulipifera*), American chestnut, white ash, American elm, ironwood (*Carpinus caroliniana*), sassafras (*Sassafras albidum*) and flowering dogwood (*Cornus florida*).

Fifteen common spring wildflowers have approximately the same distribution: blunt-lobed hepatica (*Hepatica americana*), bloodroot (*Sanguinaria canadensis*), Dutchman's breeches (*Dicentra cuccularia*), trout lily (*Erythronium americanum*), cut-leaved toothwort (*Dentaria laciniata*), wild ginger (*Asarum canadense*), jack-in-the-pulpit (*Arisaema triphyllum*), early meadow rue (*Thalictrum dioicum*), Virginia waterleaf (*Hydrophyllum virginianum*), wild geranium (*Geranium maculatum*), downy yellow violet (*Viola pubescens*) and rue anemone (*Anemonella thalictroides*) (Gleason & Cronquist, 1964) and large-flowered bellwort (*Uvularia grandiflora*). The shrub leatherwood (*Dirca palustris*) has a similar distribution.

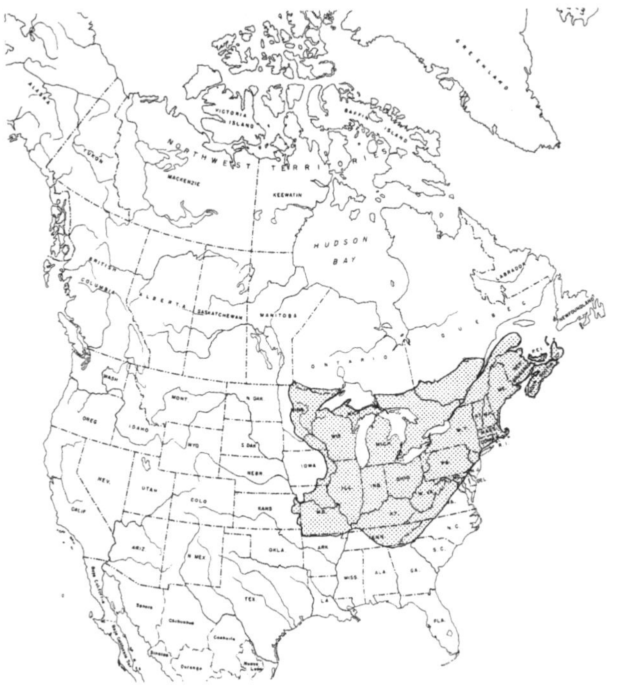

Figure 5. Distribution of sugar maple *(Acer saccharum)* (From Little, 1971)

Rare species of this widespread element are so because they may require a specialized habitat, be under pressure from picking and harvesting, or suffer from habitat elimination. Hitchcock's sedge (*Carex hitchcockiana*) and hairy woodmint (*Blephilia hirsuta*) require a rich mesic calcareous soil. Narrow-leaved spring beauty needs rich open alluvial soil. Downy woodmint (*Blephilia ciliata*) inhabits dry calcareous open ledges. The habitat of autumn coralroot (*Corallorhiza odontorhiza*), dry rich open woods, is frequent, but the small orchid is very scarce and also inconspicuous. Certain species need wet open soil, like adder's tongue fern (*Ophioglossum vulgatum*). These are early successional sites that continually are created and then become overgrown. Ginseng (*Panax quinquefolius*), requiring rich sites, may have been more common until it became a valuable commodity.

A sub-group of the Eastern Deciduous Forest, from the point of view of Berkshire County flora, has a distinctly southern and midwestern center (Figure 6.). Many of these species are the more uncommon or rare species of the county as they are at or near the northeastern limits of their ranges.

Examples of these species are: bur oak (*Quercus macrocarpa*), yellow oak (*Q. muhlenbergii*), dwarf chinquapin oak (*Q. prinoides*), bladdernut (*Staphylea trifolia*), hackberry (*Celtis occidentalis*), and prickly-ash (*Zanthoxylum americanum*). A rare species, mountain winterberry (*Ilex

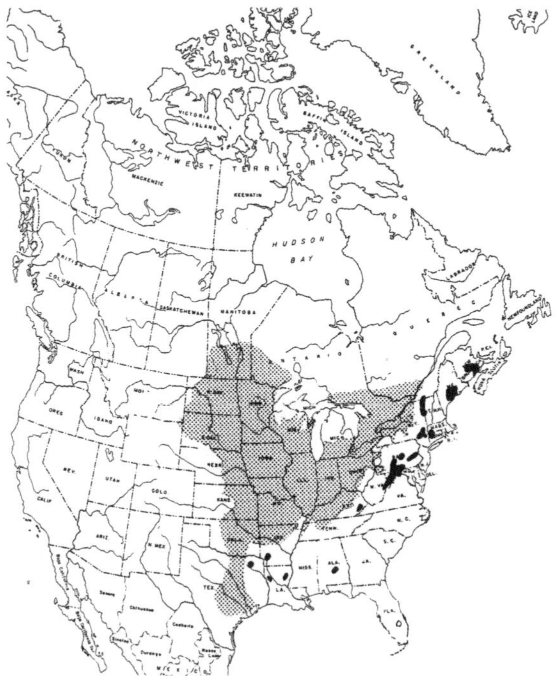

Figure 6. Distribution of bur oak *(Quercus macrocarpa)* (From Little, 1971)

Figure 7. Distribution of yellow birch *(Betula alleghaniensis)* (From Little, 1971)

montana), is at its northeastern limits in the southwest corner of the county. Its habitat, rocky, acidic woods, is common. The species may be limited by climate and the competition of better adapted species or it may still be migrating northward , still responding to climate warming after glaciation.

Certain of these species have outlying discontinuous distributions, such as bur oak. They may have expanded their ranges during the warmer, drier Hypsithermal Period, 5,000 - 4,000 years BP, and then retreated as climatic conditions changed, stranding isolated populations (Davis, 1965).

North-south-trending mountain ridges may impede the further migration eastward of certain species that are at the eastern limit of their range by harsher climate and unsuitable soil. The rare crooked-stem aster (*Aster prenanthoides*) favors alluvial soil along two rivers, the Green River in Great Barrington, that rises in New York State, and the Green River in Williamstown, which rises on the east side of the Taconic Range. It is fairly common on the western flank of the Taconics but has not found its way over the Berkshire Plateau, which has no through-flowing rivers with appropriate soil. Hairy honeysuckle (*Lonicera hirsuta*) whose main population center is in New York State and around the Great Lakes, has three stations in Berkshire County. It needs calcareous soil on open marble ledges and in open woods, which is not available on the Berkshire Plateau and further east in Massachusetts. Many other calciphilic species are similarly limited.

Alleghanian Element.

Many species, including some of our important forest trees, have a distribution limited to a southern extension along the Appalachians, widening out to an area around the Great Lakes, southern Canada and the northeast (Figure 7). Except for the Appalachian extension, it is limited to formerly glaciated regions. These species may have survived glaciation on the coastal plain, which was extensive due to lowered sea level. These pioneer species were in a position to move quickly into the poorly developed soil left by the retreating glacier. Soil south of the glacier was undisturbed, although occupied by northern conifers such as jack pine during glaciation. Once the climate warmed the Eastern Deciduous Element moved north rapidly and presented an already closed community to the Alleghanian plants spreading out from their refugia.

Most of our conifers are of this element. White pine and eastern hemlock are important members of the forests in Berkshire County. Yellow birch joins with Eastern Deciduous Forest members sugar maple and American beech to form the Northern Hardwoods Forest community. Native red pine (*Pinus resinosa*) is found only on a few open rocky ridges. Red spruce (*Picea rubens*) occurs with Boreal Element balsam fir to form the Northern Conifer Forest at high elevations. Typical of this group, staghorn sumac (*Rhus typhina*) is a vigorous pioneer species. Northern white cedar (*Thuja occidentalis*), rare as a native, appears to need calcareous soil, and may have had its refugium only in the Appalachians. Similarly, the herbaceous vine purple clematis (*Clematis*

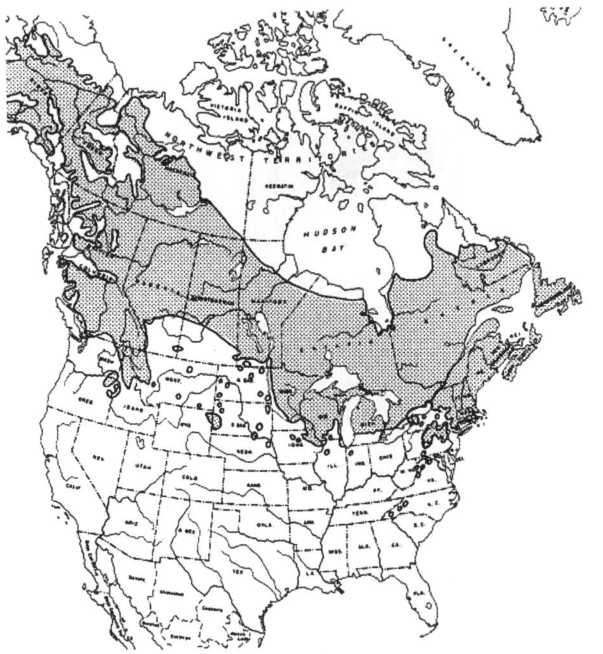

Figure 8. Distribution of paper birch *(Betula papyrifera)* (From Little, 1971)

occidentalis), may have survived in the Appalachians and the Ozarks.

<u>Boreal</u> <u>or</u> <u>Northern</u> <u>Element</u>.

This element contributes some important common species as well as rarer ones. Our common paper birch, although possessing a somewhat Alleghanian distribution southwards in the Appalachian Mountains, ranges far to the north where it spans the continent (Figure 8). It is found in almost every forest type while heart-leaved birch (*Betula papyrifera* var. *cordifolia*) is only found in high elevation Northern Conifer Forest, also called montane boreal forest, with balsam fir and red spruce. Larch or tamarack (*Larix laricina*) and black spruce (*Picea mariana*) inhabit cold bogs and fens.

Many Boreal plants have a wide Circumpolar distribution (Hulten, 1970), such as fireweed (*Epilobium angustifolium*), dwarf enchanter's nightshade (*Circaea alpina*), leatherleaf (*Chamaedaphne calyculata*), twinflower (*Linnaea borealis*), buckbean (*Menyanthes trifoliata*) and Selkirk's violet (*Viola selkirkii*).

As the late Pleistocene climate warmed, and the Wisconsinan glacier retreated, Boreal species, once widespread, were displaced by species adapted to warmer climates. They migrated north or upward in elevation, resulting in some Boreal species being stranded as disjuncts (plants that are far away from the main populations) in cold ravines, cold bogs and mountain tops in Berkshire County. Mountain cranberry exists only as one plant on a high exposed ledge. Braun's holly fern (*Polystichum braunii*) inhabits deep cold ravines where there is a calcareous influence. Smooth woodsia (*Woodsia glabella*) is found on only one high cliff, as is northern prickly rose (*Rosa acicularis*). The existence of these plants is tenuous, since they occupy small habitat "islands" where they are subject to chance extinction. Once gone, they probably would never return, since seed and spore sources are now far to the north.

Other Boreal species are near their southern limits, such as Bartram's shadbush (*Amelanchier bartramiana*) and northern mountain-ash (*Sorbus decora*). Many Boreal or circumpolar species do not inhabit strictly boreal habitats, but are found in the mix of other elements that characterizes our flora.

<u>Coastal</u> <u>Plain</u> <u>Element</u>.

During glacial periods, a great deal of water was tied up as ice, causing much lower sea levels and exposing the wide Continental Shelf. This area served as a refugium for many species (Davis, 1965), some of which remained centered on the coastal plain after the glacier disappeared.

The few species of this element are a less common part of our flora. Yellow-eyed grass (*Xyris difformis*), small floating bladderwort (*Utricularia radiata*), Massachusetts fern

Figure 9. Distribution of golden club *(Orontium aquaticum)* (from Grear, 1966.

(*Thelypteris simulata*), lance-leaved violet (*Viola lanceolata*), and bayonet rush (*Juncus militaris*) are found in wet sands, bogs or boggy ponds, acidic marshes and shallow acidic ponds typical of the coastal plain.

Golden club (*Orontium aquaticum*) (Figure 9) exists as one plant in one sandy-bottomed pond and may have disappeared as of this writing, as it has from the few other western Massachusetts ponds it formerly inhabited, which are clustered in the Connecticut River drainage. The plant probably expanded north along Connecticut River tributaries, and now, through human-assisted disturbance, is retreating to its original coastal plain habitat.

Alien Element.

Approximately 27% of the flora is alien or not in digenous to the area. Seventeen percent of the flora in Hoffmann's time (1922) was alien. Most alien species originated in Europe and Asia, but a few are North American, having taken advantage of massive land clearing and soil disturbance to expand their range.

Many were introduced very early in the European settlement purposely for pasture forage such as timothy (*Phleum pratense*), orchard grass (*Dactylis glomerata*); for herbal use such as thyme (*Thymus serpyllum*) and wild marjoram (*Origanum vulgare*); for food such as wild parsnip (*Pastinaca sativa*) and wild carrot (*Daucus carota*), and many others for decorative use.

Others arrived by chance in grains and seeds, raw wool, packing, ballast, clothing, railroads, automobiles and food shipments. These species are successful because they had already adapted to human disturbance in their native habitat and arrive free from the disease, insects and competition they are subject to in their native land. They generally occupy disturbed habitats such as roadsides, abandoned fields, gardens, vacant lots and construction sites.

A group of aggressive alien shrubs, which were not listed by Hoffmann, or listed as uncommon, have become threats to the natural plant communities of the county. These were purposely introduced to enhance wildlife food and cover, provide a fast-growing screening or erosion control, to be a "living fence" or provide decorative berries. Common buckthorn (*Rhamnus cathartica*) spreads into old fields and has invaded open woods and calcareous ledges. Shining buckthorn (*R. frangula*), a more recent arrival, is taking over beaver marshes, wet meadows and even apprears in bogs. Morrow's honeysuckle (*Lonicera morrowii*) takes over floodplains, moist woods and roadsides. Autumn and Russian olive (*Elaeagnus umbellata, E. angustifolia*), particularly the former, take over old fields, preventing native plants from re-establishing. Multiflora rose (*Rosa multiflora*), widely planted as a "living fence" has turned into a "living hell" as it invades pastures, open woods, and valuable native habitat such as ledges. Oriental or round-leaved bittersweet (*Celastrus orbiculatus*) was introduced because it had a more abundant crop of decorative berries than the uncommon native, climbing bittersweet (*C. scandens*). Its destructive habit of swarming over shrubs and trees, effectively strangling them, has long range effects on regeneration of forests. A vine with similar tendencies was recently noticed in the Lenox-Stockbridge area; this turned out to be hardy kiwi (*Actinidia arguta*), a pretty Asian species probably escaped from one of the many estates in the area. The lesson that should emerge from these events is plant native plants if at all possible, and don't plant alien species with unknown or aggressive qualities. Most commonly used garden plants pose no danger.

New species are still arriving and spreading rapidly, at least at first. Those that Hoffmann never saw include the showy purple balsam (*Impatiens glandulifera*) spreading in marshes along the Housatonic River, yellow rattle (*Rhinanthus minor*) now appearing in wet fields, eyebright (*Euphrasia nemorosa*) in wet gravels, and aggressive aquatic species.

Two aliens that threaten the integrity of our marshes are phragmites or common reed (*Phragmites australis*) and purple loosestrife (*Lythrum salicaria*). Phragmites is now thought to be native to brackish marshes in North America, but has spread rapidly into inland marshes transported by road and utility corridor construction mainly. Purple loosestrife invades and takes over marshes and wet meadows. Control is almost impossible and is feasible only in very valuable rare habitats.

Several species, native to the midwest or western North America have expanded eastward in response to clearing of forests. Black-eyed susan (*Rudbeckia hirta* var. *pulcherrima*) is a common weed in pastures and beard-tongue (*Penstemon digitalis*) inhabits wet meadows.

Many other aliens give color to local roadsides and fields, and have been accepted as a harmless addition to the flora. Given time, control factors may evolve that keep the harmful aliens in check, but it may be too late for some native plants. And new aggressive species may be arriving if we do not place controls on introductions. The greatest danger to our native plant communities is not direct destruction by human activities, but by aggressive alien species.

6. Impact of Human Beings

Pre-settlement.

Artifacts of early Native Americans have recently been found indicating that they were here at least 6500 years BP (Mass. Historical Commission, 1993) and it appears that they camped, hunted and possibly resided here up until the arrival of European settlers. They may have followed the large game animals, such as mastodon, arriving here about 11,000 years BP. Therefore, while our flora has always been impacted by humans to some extent, massive changes did not take place until European settlers moved in from the eastern part of the state.

Two small bands of Mahicans subsisted along the alluvial plains of the Housatonic River in 1735 under their chiefs Konkapot and Umpachene. They grew corn, beans, and squash and hunted over a large territory of western Massachusetts and eastern New York. Native American populations had already been reduced by European diseases, notably smallpox (Cronon, 1983; Burns and Stevens (1988). They may have done some burning, as tribes in the eastern part of the state had done, to encourage young growth for game. There is little evidence that the northern section of the county was used for more than hunting and passage. The northern hardwoods forest does not burn easily, and probably little was done (Saterson, 1977). Large fires were maintained for heat in winter, so some cutting near settlements took place.

Although the land was generally forested, with a mosaic of different forest types covering most of the area, there were two types of open meadows. Before the rivers were dammed and controlled, flooding and ice scouring created river meadows sometimes called "intervales", which were useful agriculturally. Beavers, before they were extirpated in the late 18th century, were a major force on the landscape. They created large ponds and marshes that trapped silt and organic matter. Not only were these rich in wildlife resources, but when the beaver abandoned the dams, the ponds eventually drained, leaving rich soil ideal for grasslands, which were recognized by colonists as excellent hay land that did not need clearing (Cronon, 1983). The common notion has held that the original vegetation was a uniform blanket of forest, deep dark and claustrophobic. Rather, it was a mosaic of forests in various stages of maturity due to fire, windstorm, old age and other natural catastrophes, abundant beaver-created habitat of marshes, meadows, cut areas, edge habitats and shrub swamps, bogs, ponds, lakes, bare mountain tops, river meadows, floodplains and oxbows in various stages of filling in. Of course, the great percentage of the land was forested.

Forest Composition at Time of Settlement.

A general idea of forest composition can be gained from studying early surveying records of the 18th century which used trees as "bearing trees" and corners. Saterson looked at records for Williamstown, and found that while species composition was the same as in modern forests, distribution and frequency were different. American beech was more frequent in early forest, while maples and birches now are more common. Red oak is more frequent at all elevations at present, whereas it was restricted to the northeastern section of town, noted for its dry south-facing slopes.White pine remains concentrated on sandy outwash northeast of the Hoosic River, although it has colonized many old fields. Hemlock was a small percentage of the forest, then and now.

Land Use after Settlement.

The patterns of land use that developed in Berkshire County are typical of what happened all over New England. Permanent villages were established with land owned by individuals, unlike the temporary villages of the Native Americans and their custom of communal ownership. Initially, clearing was for agriculture and local use of timber, but as soon as commercial uses for forest products arose, intense pressure was put on the forests, and most were decimated by the end of the 19th century. Natural regeneration began early and the resulting mosaic of different stages of growth persist to this day.

Clearing progressed rapidly, and in 1799, Dwight (1822) noted clearings high on mountain ridges. By 1800, 77% of Williamstown was cleared for agriculture (Brooks, 1953). Towns on the Berkshire Plateau were not as extensively cleared, but logging proceeded.

The other major use of wood was for heating, where it was burned wastefully in open fireplaces. One household might burn 30-40 cords, representing one acre of timber, in one year. Many parts of New England experienced wood shortages by the end of the 18th century (Cronon, 1983).

Industrial Uses of the Forest.

At first the forest resources of New England, and the nation, were thought to be endless. In addition to cutting timber for building materials, many other uses for wood developed that were extremely consumptive. This trend persisted until either the forests had been cut several times or changing economic competion reduced demand.

Whipple stated that in 1828 there were 185 sawmills in the county and 38 tanneries that used hemlock bark for tanning hides. Iron furnaces in North Adams, Richmond, Cheshire, Pittsfield, Lee, Lenox, Great Barrington, Lanesboro and over the line in New York and Connecticut consumed vast quantities of charcoal. A history of Egremont (1976) describes whole mountains being "coaled off" in a season. This industry disappeared when iron sources were discovered in western North America.

Early railroad engines consumed wood. The first line was laid through Washington in 1841, and trains "wooded up" at that town. A history of Washington (Crane, 1918) noted that a trip from Springfield, a distance of 40 miles (64 km) burned 3 cords of wood, and it was estimated that in one year an engine burned 4-5 million cords of wood. Fortunately, coal had put the woodburners out of business by 1876.

The burning of calcareous rock in limekilns to produce lime, which occurred wherever there were marble deposits, put a further strain on forest resources. The only evidence remaining of this industry are numerous roads named Lime Kiln Road and some old stone buildings.

A new use of wood developed in 1867 when paper mills in Lee began using wood pulp, instead of rags, to make paper. Poplar was the preferred wood, which probably had become common after clear-cutting and farm abandonment. This source also became scarce, wood having to be "drawn" from increasingly distant regions (Federal Writer's Project,

1938). Spruce took its place. When supplies were exhausted, mills moved north or west to new sources.

The rapid destruction of the forests and the accompanying erosion, fires and landslides, did however move some county residents to start setting aside land for preservation and public enjoyment, starting with Mount Greylock, Monument Mountain and Mount Everett, in the late 19th century.

Farm Abandonment and Reforestation.

Almost as soon as the land was cleared, some of it was abandoned. Farmers left the marginal rocky farms on higher elevations to move further west to richer soil made accessible by the opening of the Erie Canal in 1825, which also could bring agricultural products back to the east. The resulting second growth forests that grew back were cut for timber as soon as they reached maturity, by around 1890 (Fisher, 1933).

After 1850 and the opening of the west, farming continued to decline. Toward the end of the 19th century, demand for wood decreased and little marketable timber remained. Fisher, in his assessment of forests in New England in 1933, described degraded, weedy, diseased forests on exhausted land. In the 1930s, the State Department of Environmental Management bought up many abandoned farms and cut-over areas, primarily on the Berkshire Plateau and other mountainous regions. On these lands the Civilian Conservation Corps (CCC) carried out extensive reforestation and created recreation areas. They planted hundreds of acres to white spruce, red spruce, Norway spruce (*Picea abies*) and red pine (Drew, 1982). The CCC dammed bogs and marshes to make ponds. Egler (1940), who was studying the region, noted, with some dismay, the alteration of natural habitats.

Old Growth Forests.

The remoteness and ruggedness of the county has helped it remain less settled. Trees on steep slopes and in inaccessible ravines escaped the saw, and a few old growth forests remain, which have been studied by Peter Dunwiddie (1993) with MNHESP help. Robert Leverett has been instrumental in discovering and documenting old growth forests (Conuel, 1994), in Western Massachusetts, and throughout Eastern United States. Old-growth stands of red spruce and Eastern hemlock form patches on the slopes of Mount Greylock. On steep slopes along the Cold River, sugar maple, yellow birch, American beech, white pine, Eastern hemlock and red spruce form one of the largest old growth forests known in Massachusetts (UMass, 1992). Tree ages range from 200 to over 400 years old. Small patches of old growth Eastern hemlock are found in Mt. Washington, and the largest may be 600 years old. Forests that were set aside in the 19th century in the "Hopper" on Mount Greylock are now approaching a mature, old growth stage that is quite magnificent.

Present Conditions.

At present, percentage in forest varies from town to town, with 30% of Pittsfield forested to 85% in Florida. The average for the county is 70%. Twenty-five percent of the county is now under some form of protection, either federal, state, town, private non-profit organizations or individuals.

Threats to habitats now are not deforestation, but large-scale development, the sprawl of roads, shopping centers and malls, pipelines and condominiums, that eat up open space around urban centers and fragment the land. Farms are declining, and the fields are reverting to forest or being developed. Changes in farm practices may affect pasture or meadow species. Instead of expanses of pasture, hayfields are managed intensively. Heavy use of fertilizer and herbicide may affect vegetation also. Increased numbers of introduced species, previously discussed, have impacted old fields and natural areas near cultivation.

Shrub and herbaceous species associated with forested habitat are abundant again. Most of the rare plants of that habitat are still extant, but rare species from early successional, naturally disturbed habitats and sandplain habitats have declined in number of species and abundance (Sorrie, 1989).

7. Summary of Factors Affecting the Flora

We have described the physical factors, the topography, geology and climate of the area, that sets the scene for and controls the vegetation. The origins of the flora have been traced from early Tertiary times to when remnants of this early flora migrated north again to colonize the "clean slate" of newly deglaciated land, moving into favorable habitats, creating soil and continuing to modify habitats. Impact of Native Americans was minimal, but post-settlement land use had and continues to have a massive effect on the pristine vegetation.

II. Habitats and Plant Communities

A habitat consists of a combination of the physical characteristics of the site. The following are some of the many factors to be considered when looking at a habitat. Topographical location: Is it on a windblown cold mountaintop or deep in a valley? Water supply: Does the rain percolate quickly into the soil leaving it dry, or is there a high water table keeping soil moist? Aspect: Is it on a sun-heated south-facing slope or on a cold northern exposure? Soil: Is the soil derived from acidic rock or calcareous? Is it rocky, sandy or formed from clay? And to complicate matters, once a plant has colonized a habitat, it will proceed to change it by adding its leaf litter to the soil, and providing shelter and shade. It then becomes difficult to separate the plant from the habitat.

Characteristics of a habitat will control what plant species grow there. A silver maple will not be growing on a quartzite mountain top, nor will wall-rue (*Asplenium ruta-muraria*) be found in crevices of a quartzite cliff. Many plant species are more tolerant of conditions and grow in a variety of habitats. After looking at many assemblages of plants in natural surroundings, one begins to see the same group of species inhabiting the same habitat over and over in many different places. Then one can begin to put into order, or classify, plant communities by the types and species of plants which are most characteristic. Given the broad variability of nature, there are no neat pigeon holes into which one can stuff plant communities. There is a great deal of overlap.

Each individual plant in a community competes with other plants for the resources of light, nutrients and water, but there usually is little interdependence between plant species. Plants through their leaf litter will affect the quality of the soil, and can create a sheltered environment, as in a forest, for other plants. However, the demise of the American chestnut possibly did not affect the adjacent oaks, except for leaving more space for them, but wildlife certainly felt the loss of the nutritious nuts. There is more interdependence between the animals, the insects, birds and mammals, and the plants. These are complex interactions which cannot be considered here.

The following plant communities are described according to the physical characteristics of the habitat, the type of vegetation (whether forest or marsh, for example) and the species that occupy the habitat, with a brief discussion on how the habitat impacts the plant species. The classification and names are based on Rawinski's (1983) classification for New England. Reference is made to Reschke's (1990) classification of ecological communities for New York State which is similar in approach.

1. Forests and Woodlands

Mesic Northern Conifer Forest Community

The montane spruce-fir forest (Cogbill, 1987), the mountain spruce-fir forest (Reschke, 1990), the boreal forest (Siccama, 1974), or the northeastern spruce-fir forest of Kuchler (1964) are alternate terms for this community dominated by red spruce and balsam fir. It is best developed above 3000′ (914 m) on Mount Greylock summit and along the ridge of Saddleball Mountain where these two species occur with heart-leaved birch and yellow birch. The summit, although drastically altered now, was described in 1799 by Dwight (1822) as being densely covered with stunted, laterally-spreading balsam fir. Other associated small trees and shrubs are American mountain-ash (*Sorbus americana*), northern mountain-ash (*S. decora*), Bartram's shadbush (*Amelanchier bartramiana*), mountain holly (*Nemopanthus mucronata*), creeping snowberry (*Gaultheria hispidula*) and hobblebush (*Viburnum lantanoides*). Some herbaceous species are skunk currant (*Ribes glandulosum*), wood sorrel (*Oxalis acetosella*), large-leaved goldenrod (*Solidago macrophylla*), bunchberry (*Cornus canadensis*), black-fruited woodrush (*Luzula parviflora* var. *melanocarpa*), and mountain woodfern (*Dryopteris campyloptera*). The ground is densely covered with shining clubmoss (*Huperzia lucidula*), wood sorrel and mosses.

High winds, shorter frost-free season, colder average temperatures, ice and frost accumulation, lower light levels, and thin soil combine to make high elevations a less favorable place for plant species. No official temperature records for Mount Greylock summit exist, but in 1838 students at Williams College recorded temperatures at the summit (3505′, 1064m) and in Williamstown (689′, 210m) simultaneously every four hours (Monthly Traveler, 1838). Results show an average of 9.2^{o} F cooler on the summit which agrees well with the estimated decrease of 1° F/328′ of elevation rise. Siccama noted 93 frost-free days. He suggested that the elevational limit of low-hanging clouds corresponds to the location of the transition zone between the boreal forest and the northern hardwoods forest. Clouds provide moisture but also reduce sunlight, and in winter, form heavy frost on trees. High winds drive ice crystals into the bark exposed above snow, and creates the "table tree" formation (Marchand, 1987) seen on the summit.

Evergreen needle-leaved trees have an advantage over deciduous trees under these conditions. They may begin to photosynthesize before deciduous trees expand their leaves and continue into early days of winter whenever temperatures are above 45°. Their flexible branches withstand heavy ice more readily.

Mesic Northern Hardwood Forest Community

A widespread community, it covers the slopes and summits of all but the highest mountains of the northern half of the county, generally speaking. Within this community, others occur on edaphically, topographically or climatically specialized areas. Upper elevational limit of this forest is from

2598′-2788′ (792-850m). Towards the southern part of the county it grades into the Mesic Acidic Oak/Conifer Forest.

Sugar maple, beech and yellow birch are dominant, occurring in varying proportions throughout the community, with yellow birch becoming dominant at higher altitudes. White ash, black birch, paper birch, red oak, black cherry and red maple commonly occur in this forest, depending on soils and succession from previous disturbances. Red oak is a permanent part of this forest on well-drained south-facing slopes where the forest is more open. Otherwise it cannot regenerate in the shade of the three dominant trees. Eastern hemlock is most often seen in patches or pure stands on steep north-facing slopes and ravines, and moist stream banks. It is more common on the Berkshire Plateau where the Mesic Northern Hardwood Forest mixes with spruce and fir, and as part of the Mesic Acidic Oak/Conifer Forest. Egler (1940) described the forest of southern Berkshire Plateau as being dominated by beech and hemlock, with sugar maple and yellow birch being less common.

Striped maple (*Acer pensylvanicum*), mountain maple (*A. spicatum*), hop hornbeam (*Ostrya virginiana*), and witch hazel (*Hamamelis virginiana*) form the understory. Shrubs include American yew (*Taxus canadensis*), hobblebush, maple-leaved viburnum (*Viburnum acerifolium*), red-berried elder (*Sambucus pubens*), and American fly honeysuckle. The vernal flora is somewhat limited, but spring beauty (*Claytonia caroliniana*), trout lily (*Erythronium americanum*), early yellow violet (*Viola rotundifolia*) and red trillium (*Trillium erectum*) are widespread. Intermediate woodfern (*Dryopteris intermedia*) and Christmas fern (*Polystichum acrostichoides*) are abundant, along with shining clubmoss, partridge-berry (*Mitchella repens*), whorled wood aster (*Aster acuminatus*) and white wood aster (*A. divaricatus*).

Factors that affect the extent of this forest are climate and soil. Siccama (1974) found the frost-free season in Green Mountain Northern Hardwoods to average about 144 days. Generally the soil is moist but well-drained and more often derived from schist, less so from gneiss or quartzite. In studies of New Hampshire forests, Leak (1978) found typical Northern Hardwoods on fine till. Kudish indicated that sugar maple and other deciduous trees needed at least a 12-36″ (30-90 cm) layer of till to grow well, while conifers can survive on 3-6″ (8-15 cm) of till. His studies in the Catskills showed that till depth was more limiting than pH or elevation, although elevation in Berkshire County is certainly a limiting factor. It is difficult to say why this forest is not widespread in south county; less mountainous topography and therefore a more disturbed landscape and possibly a drier slightly warmer climate may play a part.

Rich Mesic Forest Community

Occurring in smaller, discontinuous sites throughout the county within other forest communities, this forest is found associated with calcareous bedrock and accompanying alkaline ground water. These sites tend to be the low hills edging the valleys, or lower slopes of calcareous (Stockbridge Formation) hills capped with schist or quartzite, but also high on Mount Greylock's slopes wherever calcareous rock outcrops (Walloomsac Formation).

Sugar maple is the dominant tree, sometimes to the near exclusion of other tree species, but the other common associates are white ash, American basswood and bitternut hickory (*Carya cordiformis*). Beech, black birch and red oak also occur. Red oak thrives on these sites, but does not regenerate well in the shade of sugar maple and beech. Hop hornbeam is abundant as an understory tree, and may reach canopy size occasionally. Leatherwood (*Dirca palustris*), an early spring-flowering shrub, is a good indicator of rich conditions.

Herbaceous species include a rich vernal flora, abundant ferns, grasses, sedges and other ground layer species and are listed in approximate order of decreasing degree of association with the commuunity.

Herbaceous species of the Rich Mesic Forest

Blue cohosh *Caulophyllum thalictroides*
Maidenhair fern *Adiantum pedatum*
Sharp-lobed hepatica *Hepatica acutiloba*
Wild leek *Allium tricoccum*
Wild ginger *Asarum canadense*
Toothwort *Dentaria diphylla*
Cut-leaved toothwort *Dentaria laciniata*
Squirrel corn *Dicentra canadense*
Dutchman's breeches *Dicentra cucullaria*
Large-flowered bellwort *Uvularia grandiflora*
Goldie's fern *Dryopteris goldiana*
Narrow-leaved spleenwort *Diplazium pycnocarpon*
Broad-leaved waterleaf *Hydrophyllum canadense*
Plantain-leaved sedge *Carex plantaginea*
Hitchcock's sedge *Carex hitchcockiana*
Long-spurred violet *Viola rostrata*
Showy orchis *Galearis spectabilis*
Yellow ladyslipper *Cypripedium calceolus*
Millet grass *Milium effusum*
Hairy woodmint *Blephilia hirsuta*
Ginseng *Panax quinquefolius*
Black mountain-rice *Oryzopsis racemosa*
Bottlebrush grass *Hystrix patula*
Purple false melic grass *Schizachne purpurascens*
Dewey's sedge *Carex deweyana*
Stiff sedge *Carex amphibola* var. *rigida*
Sanicle *Sanicula canadensis*
Sanicle *Sanicula trifoliata*
Barren strawberry *Waldsteinia fragarioides*
Early meadow rue *Thalictrum dioicum*
Rue anemone *Anemonella (Thalictrum) thalictroides*
Kidney-leaved buttercup *Ranunculus abortivus*

Virginia waterleaf *Hydrophyllum virginianum*
Rattlesnake fern *Botrychium virginianum*
Canada violet *Viola canadensis*
Zigzag goldenrod *Solidago flexicaulis*
Pubescent sedge *Carex hirtifolia*
Bloodroot *Sanguinaria canadensis*
Dolls' eyes *Actaea pachypoda*
Baneberry *Actaea rubra*
Wood nettle *Laportea canadensis*
Sweet cicely *Osmorhiza claytonii*

Soil in this habitat is derived from the soft, easily weathered calcareous bedrock. High pH facilitates microbial activity and the decomposition of leaf litter, resulting in a highly fertile soil with much humus. Leaves of sugar maple are rich in bases (Dickinson & Pugh, 1974) which also favor rapid decomposition and recycling of nutrients in the soil. The right amount of moisture is also crucial to the formation of these mull soils (Bormann & Buell, 1964). Because of rapid decomposition, litter layer is thin compared to that of an oak forest, making it possible for the small vernal species to receive adequate light so that they may complete their life cycle before the canopy leafs out. High fertility also promotes this process.

Mesic Acidic Oak/Conifer Forest Community

This community is more commonly found in the southern half of the county on rocky acidic soil on lower slopes. North Mountain in Cheshire, east slope of Lenox Mountain and Stone Hill in Williamstown have good examples. It is a difficult community to characterize, as it grades into the Mesic Northern Hardwood Forest and Dry Acidic Oak/Conifer Forest, or represents a successional stage after disturbance (McVaugh, 1958). Kuchler's mapping of this area as transitional between Northern Hardwood Forest and the more southerly Appalachian Oak Forest indicates the variability of this community.

Red oak, red maple and Eastern hemlock share dominance in this forest community. White pine and American beech can be important components. Sugar maple and yellow birch are scarce. Black cherry, black birch and paper birch are common. Understory trees and shrubs include striped maple, shadbush (*Amelanchier arborea*), witch hazel, mountain laurel (*Kalmia latifolia*) and maple-leaved viburnum. The herbaceous ground layer is limited, and includes clubmosses, Christmas fern, rock polypody (*Polypodium appalachianum*), Indian cucumber (*Medeola virginiana*), Canada mayflower (*Maianthemum canadense*), white wood aster, fringed bindweed (*Polygonum cilinode*), Pennsylvania sedge (*Carex pensylvanica*), and Wild sarsaparilla (*Aralia nudicaulis*).

Soil derived from acidic bedrock, such as quartzite or gneiss and often schist, tends to be thin, rocky, well-drained and poor in nutrients. This seems to be an important factor in the distribution of this community, as Kudish states that characteristic tree species of this commmunity can subsist on 8-18" (20-45 cm) of till depth. Leaves of oak species do not decompose quickly, resulting in a thick litter layer, and they contribute to the acidity of the soil, as do hemlock and white pine (Dickinson & Pugh, 1974). This soil limits growth of sugar maple, and favors a more open canopy which allows regeneration of oak species.

Dry Acidic Oak/Conifer Forest Community

This community, dominated by oak species, occurs on soils derived from acidic rock, typically quartzite and schist, on the upper slopes and south-facing slopes of mountains and is more common in south county. Pine Cobble Mountain in Williamstown, East Mountain in Great Barrington and southern Taconics have good examples of this forest.

Chestnut oak (*Quercus montana*), with black oak and pitch pine are the typical trees species, but white pine, paper birch, red maple and black birch are also common. American chestnut may attain small tree size before succumbing to the blight. Sassafras is locally abundant. The canopy is quite open, and the shrub layer is dominated by ericaceous species such as low bush blueberry (*Vaccinium angustifolium*), early sweet blueberry (*V. pallidum*), huckleberry (*Gaylussacia baccata*), mountain azalea (*Rhododendron prinophyllum*) and sheep laurel (*Kalmia angustifolia*). Wintergreen (*Gaultheria procumbens*) and trailing arbutus (*Epigaea repens*) are common evergreen ground cover species. Herbaceous species are few and are represented by bracken (*Pteridium aquilinum*), cow-wheat (*Melampyrum lineare*), pink ladyslipper (*Cypripedium acaule*), whorled loosestrife (*Lysimachia quadrifolia*), and, uncommonly, large whorled pogonia (*Isotria verticillata*). This community shares many species with, and grades into, the Southern Acidic Rocky Summit Community.

Because decomposition is slow in acidic soils, litter levels are high and soils tend to be thin and infertile. The open canopy and often south-facing slopes allows high insolation of the soil, warming and drying it. Succession to a more mesic and closed canopy forest may take place as soils accumulate, but this process can be set back quickly by an occasional fire which this community is particularly susceptible to because of high levels of dry litter, or drought. Pitch pine regeneration is facilitated by fire, which opens their cones and often creates the necessary bare soil for seed sites.

Dry Calcareous Oak/Conifer Forest Community

This community occupies low elevation, well-drained slopes and ridges, where calcareous rock is close or at the surface, primarily in Sheffield and Great Barrington. A few scattered examples exist in more northern parts on dry south-facing slopes but lack several of the characteristic species. These are marked with an asterisk in the list of species.

White oak, chestnut oak and yellow oak* may dominate in various proportions, while shagbark hickory (*Carya ovata*), hop hornbeam and white pine are also important. Shagbark hickory and pignut hickory, along with hop hornbeam form a distinct uncommon association, an open woodland usually on ridge tops, which often hosts several rare species including drooping speargrass (*Poa languida*)*, shining wedgegrass (*Sphenopholis nitida*)* and fairy wand (*Chamaelirium luteum*)*. The understory trees, flowering dogwood (*Cornus florida*) and bladdernut flourish under an open canopy. Herbaceous species are abundant, and are listed in order of decreasing degree of association with the community.

Herbs of the Dry Calcareous Oak/Conifer Forest

Four-leaved milkweed *Asclepias quadrifolia*
Sicklepod *Arabis canadensis*
Violet bush-clover *Lespedeza violacea**
Wand bush-clover *Lespedeza intermedia*
Carolina rose *Rosa carolina*
Smooth false-foxglove *Aureolaria flava*
Downy false-foxglove *Aureolaria virginica**
Woodland flax *Linum virginianum*
Hispid buttercup *Ranunculus hispidus* var. *hispidus**
Prostrate tick-trefoil *Desmodium rotundifolium**
Fairy wand *Chamaelirium luteum**
Stout woodland muhly *Muhlenbergia sylvatica*
Slender woodland muhly *Muhlenbergia tenuiflora**
Wild pink *Silene caroliniana**
Forked chickweed *Paronychia canadensis**
Woodland sunflower *Helianthus divaricatus**
Thread-leaved sedge *Carex eburnea*
Three-lobed violet *Viola triloba**
Harebell *Campanula rotundifolia*
Horse-gentian *Triosteum aurantiacum**
Round-lobed hepatica *Hepatica americana*
New Jersey tea *Ceanothus americana*
Naked tick-trefoil *Desmodium nudiflorum*
Panicled tick-trefoil *Desmodium paniculatum*
Perfoliate bellwort *Uvularia perfoliata*
Rattlesnake weed *Hieracium venosum*

Bedrock, slope and aspect all play an important part in the composition of this community. The soil, derived from calcareous bedrock, is thin and dry and the bedrock is near or at the surface. In north county, this community is found only on a few lower elevation, south-facing slopes where high insolation creates more evaporation and warmth, while in south county these conditions do not seem as important. Calcareous bedrock areas are more numerous there, the valleys more open and there may be enough of a difference in rainfall and temperature to favor development of this community. The high number of herbaceous ground layer species relates to calcareous soil and more open canopy of a oak-dominated forest.

Floodplain Forest Community

This easily-defined community occurs on deep alluvial soil laid down during thousands of years of rivers flooding over wide floodplains or narrow bands along rivers, depending on the topography. Included within the forest are oxbow ponds and river marshes. The lowest levels may be flooded annually, while higher terraces are flooded only in unusual events. The silt bars and mud banks provide small, pioneer habitats for a few annuals such as Frank's lovegrass (*Eragrostis frankii*) and awned flatsedge (*Cyperus aristatus*). The best examples of this community are along the Housatonic and Green Rivers, but remnants appear along the Hoosic, Williams and Konkapot Rivers. Much of this rich alluvial habitat has been cleared for agriculture or, in former years, unfortunately used as dumps.

Black willow (*Salix nigra*), sycamore and cottonwood are early colonizers of banks and silt bars (Nichols, 1916) and are found nearest the river. Box-elder (*Acer negundo*) is common on Hoosic River floodplains, while silver maple is dominant in Housatonic River Floodplain Forests. American elm was an important part of the floodplain forest until succumbing to Dutch elm disease. Green ash (*Fraxinus pennsylvanica*) is generally found on river banks, while butternut, American basswood and hackberry (*Celtis occidentalis*) occupy the upper alluvial terraces.

There are few shrubs, except for the weedy Morrow's honeysuckle, but several woody vines, riverbank grape (*Vitis riparia*), Virginia creeper (*Parthenocissus quinquefolia*) and moonseed (*Menispermum canadense*), and herbaceous vines wild cucumber (*Echinocystis lobata*) and bur cucumber (*Sicyos angulatus*) flourish.

A great many herbaceous species occur only in floodplain forests or on silt bars and banks. These are listed in approximate order of decreasing association with the community.

Herbaceous species of the Floodplain Forest Community

Green dragon *Arisaema dracontium*
Davis's sedge *Carex davisii*
Gray's sedge *Carex grayii*
Hairy-fruited sedge *Carex trichocarpa*
Tuckerman's sedge *Carex tuckermanii*
Wapato *Sagittaria cuneata* (in oxbow ponds)
Giant Solomon's seal *Polygonatum biflorum* var. *commutatum*
Wild garlic *Allium canadense*
Clotbur *Xanthium strumarium*
Giant ragweed *Ambrosia trifida*
Sneezeweed *Helenium autumnale*
Floodplain germander *Teucrium canadense* var. *virginicum*
Wild black currant *Ribes americanum*

Narrow-leaved spring beauty *Claytonia virginiana*
Foxtail sedge *Carex alopecoidea*
Greenhead coneflower *Rudbeckia laciniata*
Crooked-stem Aster *Aster prenanthoides*
Wiegand's wild-rye *Elymus canadensis* var. *wiegandii*
Riverbank wild-rye *Elymus riparia*
Hairy wild-rye *Elymus villosus*
Canada brome *Bromus altissimus*
Ostrich fern *Matteuccia struthiopteris*
Pubescent sedge *Carex hirtifolia*
Floodplain avens *Geum laciniatum*
Maryland figwort *Scrophularia marilandica*
Starry false Solomon's seal *Smilacina stellata*
Wood nettle *Laportea canadensis*
False-pimpernel *Lindernia dubia*
Marsh yellowcress *Rorippa palustris*
Sprengel's sedge *Carex sprengelii*
Stinging nettle *Urtica dioica* var. *dioica*
Jerusalem artichoke *Helianthus tuberosus*

Inundation, depth of soil and high water table are important factors in this community. Woody or herbaceous perennials must be adapted to withstand a reduced oxygen supply to their roots during floods, possibly every spring, and to tolerate a high water table most of the time. At times of extreme high water, the force of the water is considerable. Multi-stemmed shrubs would be at a disadvantage since debris would catch on their branches and they might be totally submersed. Ice scours banks and silt bars, resulting in a continually disturbed habitat open for colonization by annuals. The cutting of a new channel will move a river away from its formerly adjacent floodplains and oxbows, setting in motion a change of habitat. Rivers at present are far more controlled by hardened embankments, dams and other flood control devices, but the Hoosic and Housatonic Rivers still claim their floodplains during spring floods.

The soil, silt that is continually deposited by the river, is deep, without distinct horizons and contains considerable organic matter (Rawinski, 1983). Woody species typically grow to a large size on this fertile soil, and herbaceous growth tends to be rank and tall. Calcareous rock underlies valley regions, and probably influences the richness of the soil.

<u>Pitch Pine/Scrub Oak Barren and Associated Sandplain Grassland Communities</u>

These communities once occurred, closely associated together in the same habitat, on the sandy outwash plains along the Housatonic and Konkapot Rivers in the southernmost towns of Sheffield and New Marlborough. At present, agriculture, residential housing and sand pits fragment these natural habitats. Those left in a more or less natural state are moving toward a forest dominated by white pine and tree oaks, such as red, white and scarlet oaks (*Quercus coccinea*) with little regeneration of pitch pine. At the time of Hoffmann's investigation of the flora, 40% of Sheffield was forested, and possibly the burning that formerly was carried on by Native Americans still occurred, by accident or purposely. Now fire is suppressed and due to development may never be a factor again. Species typical of these communities exist in small clearings and roadsides.

Formerly dominant, pitch pine is being superseded by white pine, white oak, red oak, scarlet oak , red maple and black cherry. Scrub oak (*Quercus ilicifolia*) now occurs at edges of this forest. Lowbush blueberry, low sweet blueberry and huckleberry persist as the shrub layer, along with species found in Dry Acidic Oak/Conifer Forest.

The following species typically associated with the Pitch Pine/Scrub Oak and Sandplain Grassland Communities are found on roadsides or clearings in the sandplain area: Dwarf chinquapin oak (*Quercus prinoides*), shining sumac (*Rhus copallina*), sand-rush (*Bulbostylis capillaris*), short fescue sedge (*Carex brevior*), sand flatsedge (*Cyperus filiculmis*), secund rush (*Juncus secundus*), Scribner's panic grass (*Dichanthelium oligosanthes*), little bluestem (*Schizachyrium scoparium*), bastard toad-flax (*Comandra umbellata*), orange-grass (*Hypericum gentianoides*), Canadian frostweed (*Helianthemum canadense*), Pinweed (*Lechea intermedia*), Blue toad-flax (*Linaria canadensis*), stiff aster (*Aster linearifolius*), and dwarf dandelion (*Krigia virginica*). Species that were listed by Hoffmann as occasional or common on the sandplain,but have not been reported since, include Bicknell's frostweed (*Helianthemum bicknellii*), wild lupine (*Lupinus perennis*), rattlebox (*Crotalaria sagittalis*), bluntleaf milkweed (*Asclepias amplexicaulis*), and butterfly weed (*Asclepias tuberosa*).

These communities are strongly influenced by dry, sandy soil and are fire-dependent. Breakdown of litter is slow because of dryness and low pH of soil, resulting in build-up of litter, a condition conducive to fire. Fire removes litter, creating bare mineral soil, and allowing sunlight to reach the soil (Henderson, 1982), which favor the establishment of pitch pine seedlings. The absence of burning favors a vegetational change to tree oak species, other hardwoods and white pine, as noted by Seischab and Bernard (1991) in New York State.

2. Rocky Summit and Cliff Communities

These two types of communities are limited in extent, as exposed rock is rare in this generally mesic region. Calcareous rock, being soft, is easily worn away, resulting in low ridges rather than high, sharp outcrops. The two communities may grade one into another from ridge to shaded ledges and cliffs.

Southern Acidic Rocky Summit Community

Occupying mountain summits and ridges, this community is more common on the southern Taconic mountains with the elevation ranging from 1900-2500′ (579-761 m). Mount Everett, Mount Race, and Alander Mountain in the southwest portion of the county, and East and Monument Mountains in the center and Pine Cobble Mountain in the north have good examples of this community.

Vegetationally this community is closely related to the Dry Acidic Oak/Conifer Forest, but it is characterized by larger areas of open bedrock, scattered small trees and dense shrub growth. Pitch pine is always present and on the southern Taconic ridges, growth of scrub oak may be dense. Paper birch, red oak and red maple may appear as stunted trees. On Tom Ball Mountain summit, our three native pines grow together: white pine, pitch pine and the uncommon red pine which only grows in this community indigenously. A dense cover of ericaceous shrubs is formed by lowbush blueberry, huckleberry and mountain azalea. Bearberry (*Arctostaphylos uva-ursi*) occurs in open, sunny clearings on a few southern summits. Black chokeberry (*Aronia melanocarpa*) and running shadbush (*Amelanchier stolonifera*) mix with the other shrub species. Three-toothed cinquefoil (*Potentilla tridentata*) appears frequently in clearings, as it does in higher, more northern open ridges.

Diversity of herbaceous species is low. Rusty woodsia (*Woodsia ilvensis*) occurs occasionally in southern locations. Hairgrass (*Deschampsia flexuosa*) and little bluestem are common tuft-forming grasses. Pennsylvania sedge forms extensive patches. Pink corydalis (*Corydalis sempervirens*) seemingly thrives in pockets of soil in open rock.

An unusual dwarf pitch pine/scrub oak "forest" is found on the level summit of Mount Everett and on a ridge in Clarksburg without scrub oak. Trees average 3 feet (1 m) in height, and have flat-topped aspect. Keith (1912) quotes Professor Edward Hitchcock's description of Mount Everett's summit in 1839: "a naked eminence with numerous yellow pines, two or three feet high". McIntosh notes a similar dwarf pitch pine forest on a summit in the Shawangunk Mountains of New York.

Lack of nutrients, heat and dryness and possibly fire are all factors in the creation of this community. There is no layer of till on these ridges, and acidic bedrock has produced little soil. Sparse vegetation allows full sun to reach the gound, resulting in extremes of temperature and dryness. During a droughty hot July in 1993, stunted tree species of oak, red maple and beech had dropped their leaves, shrubs were brown, but small herbaceous species, adapted to such xeric conditions, were thriving. McVaugh (1958) commented on the possible succession of this community and concluded that it appeared stable. Hoffmann, in 1922, noted the same species, particularly a patch of bearberry on West Stockbridge Mountain which had probably been seen by Dewey in 1829.

Southern Calcareous Rocky Summit Community

This community occurs on ridgetops of low hills edging the valleys, where calcareous rock outcrops. The bedrock of certain ridges appears to be schist, like Tom Ball and West Stockbridge Mountains, but as shown by the vegetation, there is some calcareous influence.

These, open dry rock outcrops support vegetation that is intolerant of much shade and require calcareous soil. These species are often low-growing and are adapted to growing in crevices with little soil. Purple clematis is often in this habitat, along with another vine, hairy honeysuckle; both are rare. Two small species, false pennyroyal (*Trichostema brachiatum*) and rock sandwort (*Minuartia michauxii*), are uncommon. Thread-leaved sedge may form a dense cover on rocks. A northern element, northern prickly rose, exists on a high ridge in north county, growing with balsam ragwort (*Senecio pauperculus*). West Stockbridge Mountain supports a mixture of lime-loving and acidic species. Long-leaved bluet (*Hedyotis longifolia*) and lyre-leaved rockcress (*Arabis lyrata*) grow near rusty woodsia in crevices in open rock. Roundleaf shadbush forms small patches close to huckleberry, pitch pine and three-toothed cinquefoil. Downy arrowwood (*Viburnum rafinesquianum*) grows on more shaded slopes.

The open aspect of these ridges is caused by dry, thin soil and by the tendency of trees, once full grown, to uproot and pull away from steep slopes. Rocks splitting off from ridges also renew open areas. In the moist climate of New England, however, change to a more closed canopy is proceeding, and as McVaugh stated, this community is becoming less abundant. It is also subject to being taken over by alien weed species, such as barberries, buckthorns, multiflora rose and Oriental bittersweet.

Southern Acidic Cliff Community

The habitat available for this community is widespread throughout the county in quartzite, gneiss or schist bedrock. The vegetation is sparse and not characterized by a specific or distinctive group of plants, unlike the following community. Rock polypody (*Polypodium appalachianum*) is always present in crevices and forms large mats on boulder tops. Bristly sarsaparilla (*Aralia hispida*), marginal fern (*Dryopteris marginalis*), bush honeysuckle (*Diervilla lonicera*) and Virginia creeper are species commonly found in acidic situations. Two rare species of this community are mountain spleenwort (*Asplenium montanum*), found on quartzite ledges in Sheffield, and climbing fumitory (*Adlumia fungosa*), on boulder talus below cliffs.

This habitat is cooler, moister and more shaded than the Summit Community. The acidic rock produces little soil and nutrients for vegetation, and when shaded, becomes unfavorable for many species.

Southern Calcareous Cliff Community

This uncommon community is found on steep cliff faces or sloping rock outcrops. Frequently, the ridge is capped with resistant rock, which has kept the soft calcareous rock from being completely worn down. The habitat, while found throughout the western part of the county, is best represented in Sheffield, particularly at Bartholomew's Cobble, a Trustees of Reservations property open to the public. Here, a certain type of marble is full of solution cavities and crevices, ideal for small herbaceous species.

The small ferns, purple-stemmed cliffbrake (*Pellaea atropurpurea*), wall-rue (*Asplenium ruta-muraria*), ebony spleenwort (*Asplenium platyneuron*), maidenhair spleenwort (*Asplenium trichomanes*), and blunt-lobed woodsia (*Woodsia obtusa*) colonize sunny crevices. Walking fern (*Asplenium rhizophyllum*) covers shaded mossy rock faces. Rock spikemoss (*Selaginella rupestris*) prefers shallow soil pockets in open dry rock, as does harebell. Pellitory (*Parietaria pensylvanica*) colonizes dry to moist outcrops. Lyre-leaved rockcress, smooth rockcress (*Arabis laevigata*), hairy rockcress (*Arabis hirsuta*), and early saxifrage (*Saxifraga virginiensis*) prefer more shaded ledges, as do bulblet fern (*Cystopteris bulbifera*) and fragile fern (*Cystopteris tenuis*). Two rare northern elements are slender rock-brake (*Cryptogramma stelleri*) which inhabits dripping cliffs, and smooth woodsia (*Woodsia glabella*) which grows in crevices on a high northern ledge.

As few trees and shrubs can gain a foothold on these steep rock faces, there is sufficient light and space for these small species which are adapted for survival on little soil that occasionally dries out. The easily weathered marble rock provides alkaline soil which facilitates uptake of nutrients, and supplies many footholds for plants.

Serpentine Outcrop Community

There are five or six locations in the county for ultramafic rock, igneous rock that has high percentages of iron and magnesium, only two of which occur as exposed outcrops. These are part of a narrow belt of serpentine rock outcrops along the east flank of the Appalachian Mountains that extends from Alabama to Newfoundland (Zika & Dann, 1985). They occur as intrusive rock in quartzite, gneiss and schist on the Berkshire Plateau.

The aspect and typical species composition of vegetation on serpentine soils are in marked contrast to those of non-serpentine areas. The vegetation is characterized by open canopy of sparse and stunted trees (Walker, 1954), few ground layer species on dry bare soil, and presence of endemic or disjunct species (Zika & Dann, 1985). The outcrops in Berkshire County are not large enough to support an extensive serpentine flora. Large-leaved sandwort is the only species that is absolutely restricted to serpentine outcrops, and is found on two of the sites. Field chickweed (*Cerastium arvense*) is most abundant on serpentine outcrops, but is also found occasionally on ledges in open woods on south-facing slopes. Maidenhair spleenwort and harebell are frequent on serpentine formations (Zika & Dann, 1985). The most extensive outcrop, in the town of Florida, also supports other calciphilic species such as rock spikemoss and columbine (*Aquilegia canadensis*). Where the serpentine formation is covered by a thick layer of soil, the vegetation is not notably different. The site in Cheshire has an unusually large expanse of bristly clubmoss (*Lycopodium annotinum*).

Serpentine soils are infertile and actively inimical to plant growth, and only species adapted to those conditions can survive. The soils derived from serpentine rock are high in magnesium, iron and heavy metals chromium and nickel, but usually deficient in calcium, nitrogen and phosphorus (Walker, 1954; Brooks, 1987). The high ratio of magnesium to calcium is a limiting factor to plant growth, since calcium acts to prevent damage from magnesium and nickel. Plants adapted to serpentine soil are able to absorb calcium even if it occurs at very low levels, and to be resistant to damage from nickel. Kruckeberg (1954) showed that serpentine endemics are able to grow on normal soil, but cannot compete with other vegetation. The open canopy and bare soil provide a pioneer habitat for the endemics which is not invaded by the more shade-tolerant, non-serpentine species.

3. Lakes and Ponds

These aquatic habitats are classified according to the alkalinity of the water, which is a major controlling factor in the distribution of species in these water bodies. Hellquist (1975), in his study of the distribution of pondweeds (*Potamogeton* spp.) in New England, found that the occurrence of most pondweed species was correlated with certain ranges of alkalinity, and he classified them into four groups. Some species tolerate a wide range of alkalinity. Other aquatic species also have distributions limited to a certain range of alkalinity. Depth of water, bottom conditions and elevation are also important factors in determining species composition or plant communities of aquatic habitats.

Clear Softwater Lake/Pond Community

These water bodies occur in regions of acidic bedrock at elevations between 1469′ and 2040′ (448-622 m), mainly on the Berkshire Plateau with a few on the uplands of the Taconic Range.

The vegetation in general is characterized by submersed species with a rosette growth form and pondweeds that are restricted to very acidic waters. Spiny quillwort (*Isoetes echinospora*), pipewort (*Eriocaulon aquaticum*), grass-leaved arrowhead (*Sagittaria graminea*), waterwort (*Elatine minima*), and water lobelia (*Lobelia dortmanna*) all have submersed rosettes of leaves. Certain pondweeds (*Potamogeton bicupula-*

tus, P.confervoides, P. oaksianus) are found only in these extremely acidic waters. Floating-leaved species are the widespread yellow water-lily (*Nuphar variegata*), the infrequent floating bur-reed (*Sparganium fluctuans*), and uncommon floating-heart (*Nymphoides cordata*). Native milfoils (*Myriophyllum tenellum, M. farwellii, M. humile*) are uncommon but typical of this community.

Cooler climate, abundant moisture and acidic bedrock of these uplands are imiportant factors in this community. The water contains up to 18 mg/liter HCO_3^- (Hellquist, 1980), which is extremely low and is low in nutrient. Lake and pond bottoms are typically rocky, gravelly or sandy and therefore not suitable for deep-rooted plants. Margins are usually rocky but may have small areas of peat along the edges. Clear, shallow water is conducive to growth of submersed plants. North and South Ponds in Savoy State Forest have good examples of this community.

Acidic Brownwater Lake/Pond Community

Like the preceding habitat, this community is found at higher elevations, on the Berkshire Plateau in regions of acidic rock. It is usually associated with bogs, or boggy areas that have been altered or flooded.

As with most vegetational groupings, there is a great deal of overlap, as many species have a wide tolerance of environmental conditions. Each water body differs slightly one from another, and a wide variety of conditions such as acidity and substrate is found. The vegetation includes more species of the floating-leaved form, such as yellow water-lily, white water-lily (*Nymphaea odorata*), and watershield (*Brasenia schreberi*). Small floating bladderwort (*Utricularia radiata*) is quite common, along with common bladderwort (*Utricularia vulgaris*), which is widespread in a variety of aquatic habitats. Purple bladderwort (*Utricularia purpurea*) may occur in both the Clearwater and Brownwater habitats, while twinscape bladderwort (*Utricularia geminiscapa*) prefers bog waters. Surface pondweed (*Potamogeton epihydrus*) is fairly common in this community. Pipewort and grass-leaved arrowhead occur in the shallow margins.

Factors are similar to those of the Clearwater habitat, but depth of water, water quality and substrate differ. Alkalinity levels may range up to 30 mg/liter HCO_3^- (Hellquist, 1980). If the basin of the water body is deep, organic matter will accumulate and provide suitable substrate for deep-rooted aquatics. Lack of drainage may contribute to organic accumulation. The clarity of the water is obscured with organic acids and fine particles of organic matter, which makes growth difficult for rosette form or submersed aquatic plants, but not important to floating-leaved plants. Bog Pond in Savoy State Forest and the Spectacle Ponds in Sandisfield State Forest are good examples of this aquatic community.

Moderately Alkaline Lake/Pond Community

This aquatic community and the following are found in regions with calcareous bedrock in the Central Valley region and occur at elevation range 725-1282′ (221-391 m).

Pondweed species that more often occur within an alkalinity range of 18-73 HCO_3^- include *Potamogeton spirillus* and *P. robbinsii* as inhabiting less alkaline waters, while *P. amplifolius, P. gramineus, P. natans, P. praelongus, P. obtusifolius,* and *P. zosteriformis* are found in progressively more alkaline waters. Other aquatics that occur in moderately alkaline waters are coontail (*Ceratophyllum demersum*), common naiad (*Najas flexilis*), waterweed (*Elodea canadensis*) and occasionally *E. nuttallii*, wild celery (*Vallisneria americana*) and white water crowfoot (*Ranunculus longirostris*). Star duckweed (*Lemna trisulca*) is found more often in alkaline waters, while duckweed (*Lemna minor*) and big duckweed (*Spirodela polyrhiza*) are widespread. Northern milfoil (*Myriophyllum sibiricum*) is becoming uncommon as the introduced Eurasian milfoil takes over. Yellow and white waterlilies are found here also.

The higher alkalinity of the water, which derives from streams and springs flowing through calcareous bedrock, is an important factor in this community, as is the depth of soil. Situated mainly in the valleys, where till, outwash and alluvial soil are abundant and more or less alkaline, the ponds and lakes have accumulated thick layers of bottom muck which provide nutrients and rooting sites. High diversity of species is due to high levels of nutrients. Good examples of this community occur in Lake Garfield, which is less alkaline, and Lakes Pontoosuc and Onota in Pittsfield and Prospect Lake in Egremont, which have higher, yet moderate alkalinity levels.

Highly Alkaline Lake/Pond Community

This diverse community is similar in most respects to the preceding one and shares most of the plant species. The alkalinity levels range from 73 to more than 109 mg/liter HCO_3^- (Hellquist, 1975, 1980). It is distinguished by several pondweed species that are more typically found in these waters, namely *Potamogeton pusillus* var. *pusillus, P. foliosus,* and *P. illinoensis* are common,while *P. friesii* is rare. *P. crispus* inhabits nutrient-enriched, sometimes polluted waters. *P. pectinatus* is restricted to the most alkaline waters, those with alkalinity levels above 109 mg/liter HCO_3^-. *P. hillii* is similarly restricted, with the greatest concentration of populations, world-wide, in Berkshire County centered around Richmond (Hellquist, 1984). It is frequent locally in cool waters of beaver ponds and small streams. Water star-grass (*Heterantera dubia*) is common in many highly alkaline sites. Water marigold (*Megalodonta beckii*), which is fairly uncommon, seems to occur in this habitat, at least in Berkshire County. Tiny cowlily (*Nuphar pumila*) exists only in one oxbow pond near the Housatonic River.

Good examples of this community occur in Shaker Mill and Cranberry Ponds in West Stockbridge, Lake Buel in Monterey and Mill Pond in Egremont. Often plants in these ponds are encrusted with marl deposits.

4. Rivers, Streams and Springs

Rivers and Streams

Rivers and streams are characterized as high-, medium- or low-gradient, depending on the steepness of the slope and rapidity of the flow. Most have stretches of both fast and slow moving waters, but usually the most rapid sections are at the headwaters rising on the mountain slopes.

There are no large high-gradient rivers, but the Deerfield, Cold and Westfield Rivers, portions of the east branch of the Housatonic River and north branch of the Hoosic River are medium-sized fast-flowing rivers. The main Hoosic River is a medium-gradient river with a rocky, gravelly bed. The Housatonic and its larger tributaries, such as the Konkapot, Green and Williams Rivers are medium- to low-gradient. The Housatonic, in particular, meanders through wide floodplains with silty banks and stream bed. Throughout the county there are numerous small brooks; some are fast-moving and rocky, others are slow-flowing with silty or sandy bottoms.

The aquatic flora of these waters is limited, with the low-gradient streams hosting the greatest diversity of species. Water-starwort (*Callitriche* spp.) grows in the shallow gravelly shores of medium-gradient streams. The Housatonic and Hoosic River systems have the first and third highest alkalinities in New England, therefore it is not surprising that the aquatic plant community resembles that of alkaline lakes and ponds. Several pondweeds inhabit the river, with *Potamogeton nodosus* being most typically found in slow rivers. *P. crispus* and Eurasian milfoil, alien species, are increasing in this habitat. Coontail and waterweed occur in backwaters and oxbow ponds. Shallow waters are colonized by two bur-reeds, *Sparganium americanum* and *S. emersum*, which will often have long ribbon-like leaves in flowing water.

Springs

This community is found throughout the county wherever water wells up from bedrock and forms a small stream. Golden saxifrage (*Chrysosplenium americanum*) may colonize shaded springy areas. In sunny locations the introduced watercress (*Nasturtium officinale*) and natives American brooklime (*Veronica americana*) and American field mint (*Mentha arvensis*) are abundant.

5. Wetlands

There is a great variety of wetlands in the county, ranging from forested swamps to open graminoid wet meadows and level bogs. Many species of these communities have a wide latitude of occurrence. One wetland may be made up of a mosaic of several different communities, or vegetation will occur in zones around a central pond, as in a level bog. Elements of both bog and fen communities can occur in adjacent patches. Wetland communities reflect an infinite number of gradations from nutrient-poor acidic habitats to nutrient-rich marshes. This variability makes it difficult to classify these communities very neatly, but certain patterns do appear.

An important factor in the type of wetland community is the degree of acidity or alkalinity of the water which depends on its source from bedrock and soil. Depth of soil and organic accumulation significantly contribute to the type of vegetation observed. Drainage through a wetland, whether limited as in bogs or considerable as in marshes, is important to the amount of organic accumulation and nutrient supply (Dansereau & Segada-Vianna, 1952).

To clarify the use of different names for wetlands: Swamps are forested wetlands, shrub swamps are wetlands dominated by shrubs, and marshes are open wetlands dominated by emergent or graminoid species. Bogs and fens are considered separately. The communities are discussed in that order.

Acidic Conifer Swamp Community

This community occurs almost entirely on the Berkshire Plateau, where topography and climate are conducive to moist, cool, poorly-drained situations with acidic bedrock. Elevation ranges from 1500-2000′ (457-610 m). The swamp is surrounded by elements of Mesic Northern Hardwood Forest and Mesic Northern Conifer Forest Communities. This community occurs in poorly drained areas or on a bog mat that has sufficiently stabilized to allow growth of trees.

The dominant species are balsam fir and red spruce, with Eastern hemlock also present. The ground has a hummock and pool topography, and often a deep layer of sphagnum moss. Shrubs and herbaceous species are scarce. Among the typical shrubs and ground cover species are mountain holly (*Nemopanthus mucronata*) and small evergreen species creeping snowberry, twinflower (*Linnaea borealis*) and goldthread (*Coptis trifolia*). Smooth winterberry (*Ilex laevigata*) is occasionally present.Three-leaved false Solomon's seal (*Smilacina trifolia*) is uncommon. Bunchberry and three-seeded bog sedge are common on the mossy mat, and northern awned sedge (*Carex gynandra*), although abundant in other wetland communities, seems to be one of the few other sedges occurring here. Wolf Swamp, in New Marlboro, is an excellent example of a bog with complete closure of open water and a well-developed Acidic Conifer Swamp along the edge.

Acidic Hardwood Swamp Community

This community, which also is referred to as the Red Maple Swamp Community, is frequent in the county at lower elevations wherever acidic substrate occurs. Red maple is dominant, but never develops into the large trees seen in upland environments. Other tree species include American elm and white pine, which is never vigorous but survives on drier hummocks. Swamp white oak occurs more commonly in the southern part of the county, as does the uncommon sour gum (*Nyssa sylvatica*).

The canopy is somewhat open, and a variety of shrubs and herbaceous species are associated with this community, including winterberry (*Ilex verticillata*), nannyberry (*Viburnum lentago*), maleberry (*Lyonia lugustrina*) and occasionally mountain holly. Pussy willow (*Salix discolor*) and silky willow (*S. sericea*) occur in openings. The more common herbaceous species are (*Osmunda regalis*), cinnamon fern (*O. cinnamomea*), crested fern (*Dryopteris cristata*), sensitive fern (*Onoclea sensibilis*) and tussock sedge (*Carex stricta*).

The important factors are acidic substrate and water, and poor drainage that contributes to low nutrient supply. Few tree species can tolerate these conditions, but adaptable red maple can grow here with little competition. The shallow depressions where this community occurs do not allow much build-up of peat or organic soil.

Circumneutral Hardwood Swamp Community

Given the extensive calcareous deposits in Berkshire County, habitat for this and other communities dependent on more alkaline habitats are common throughout the western portion of the county. These swamps occur along rivers or streams and most typically are the headwaters source of streams.

Black ash (*Fraxinus nigra*) and red maple are dominants, but in south county swamp white oak, green ash and bur oak are also important. Larch may be present. Ironwood is common as an understory tree. Spicebush (*Lindera benzoin*) is abundant. The typical herbaceous species are skunk cabbage (*Symplocarpus foetidus*), swamp saxifrage (*Saxifraga pensylvanica*), False hellebore (*Veratrum viride*), square-stem goldenrod (*Solidago patula*), woodland horsetail (*Equisetum sylvaticum*), cinnamon fern, brome sedge (*Carex bromoides*), dwarf raspberry (*Rubus pubescens*) and small purple-fringed orchis (*Platanthera psycodes).* Presence of shrubby cinquefoil (*Potentilla fruticosa*) and alder-leaf buckthorn (*Rhamnus alnifolia*) indicate conditions close to that of fens, to which this community is closely related.

Important factors are the rate of flow and the pH of water flowing through the swamp to the stream, constantly supplying minerals and creating a seepage swamp (McVaugh, 1958). Circumneutral water facilitates nutrient uptake and decomposition of organic material.

Acidic Shrub Swamp Community

This community is widespread in regions where waters and soils are acidic, particularly on the Berkshire Plateau and occurs on margins of streams and poorly drained basins. Speckled alder (*Alnus incana* ssp. *rugosa*) often forms pure stands, but many shrub swamps include winterberry, meadowsweet (*Spiraea latifolia*), highbush blueberry (*Vaccinium corymbosum*), buttonbush (*Cephalanthus occidentalis*), sweet gale (*Myrica gale*), maleberry, silky willow, wild raisin (*Viburnum cassinoides*), and arrowwood. A few red maple saplings and shadbush or, on the Berkshire Plateau, red spruce or balsam fir are usually present. Herbaceous species include the vine virgin's bower (*Clematis virginiana*), sensitive fern, cinnamon fern, royal fern, fowl mannagrass (*Glyceria striata*) and tussock sedge. Purple-stemmed aster (*Aster puniceus*) and joe-pye weed (*Eupatorium maculatum*) occur in openings.

This habitat is similar to that of the Acidic Hardwood Swamp, and these shrub swamps may be transitional to a forested type, but occasional flooding and water level fluctuation may ensure that the shrub swamp will persist for a long time.

Circumneutral Shrub Swamp Community

The previous shrub swamp grades into this community, allowing for many different variations in species composition. This shrub swamp community is dominated by pussy willow, silky willow, shining willow (*Salix lucida*), silky dogwood (*Cornus amomum*) and red-osier dogwood (*Cornus sericea*). Spicebush is a good indicator of circumneutral conditions. Swamp rose (*Rosa palustris*) grows thickly at the edges of the swamp. Speckled alder will also be present, but shrubs are scattered. The vines virgin's bower and bittersweet nightshade (*Solanum dulcamara*) can be abundant. Among the herbaceous species, the most distinctive are skunk cabbage, swamp saxifrage, squarestem goldenrod and swamp milkweed (*Asclepias incarnata*). Other common species are cinnamon fern, crested fern, marsh violet (*Viola cucculata*), marsh marigold (*Caltha palustris*), swamp buttercup (*Ranunculus hispidus* var. *caricetorum*), and reed meadowgrass (*Glyceria grandis*). Scattered trees, such as elm and red maple, may occur.

As in other wetland communities, water supply and pH and topography are the main factors affecting the species composition of this community. Slowly-moving water originating from groundwater seepage or adjacent water bodies, with some calcareous influence from soil or bedrock, supplies nutrients. Open, flat margins of water bodies or shallow basins provide suitable conditions for organic accumulation. This community may be successional as described by McVaugh (1958), but extremely wet and occasional flooding conditions preclude extensive tree growth.

6. Marshes

Vegetation of marshes is characterized by tall emergent graminoid or broadleafed herbaceous species growing in water or saturated soil. Topography is generally flat, but there is adequate movement of water. The soil is a mixture of organic and mineral sediments (Dansereau & Segadas-Vianna, 1952).

Robust Emergent Marsh Community

This community is found in shallow water along streams, rivers, ponds and lakes throughout the county. It is dominated by tall emergents which are rooted in a thick layer of organic muck or more mineral soil. Waters and soil may vary from acidic to fairly alkaline, with the most common condition being intermediate, making it difficult to delineate two separate communities based on these conditions.

Cattail is typically present in pure stands, as it spreads by creeping rhizomes. Narrow-leaf cattail (*Typha angustifolia*) appears in more alkaline or saline situations, such as roadsides. Other common species, beginning with the more acidic environments, are wool-grass (*Scirpus cyperinus*), three-square (*S. pungens*), bur-reed, Small's spikesedge (*Eleocharis smallii*), purple-stem aster and the shrub meadowsweet. With increasing richness and alkalinity, soft-stem bulrush (*Scirpus tabernaemontanii*), hard-stem bulrush (*S. acutus*), arrowhead, pickerel weed (*Pontederia cordata*), arrow arum (*Peltandra virginica*), water-plantain (*Alisma* spp.), river horsetail (*Equisetum fluviatile*), marsh cinquefoil (*Potentilla palustris*), water-hemlock (*Cicuta bulbifera*), water-parsnip (*Sium suave*), bristly sedge (*Carex comosa*), lakeside sedge (*C. lacustris*), blue flag (*Iris versicolor*), large-fruited bur-reed (*Sparganium eurycarpum*) and swamp candles (*Lysimachia terrestris*) may be present.

Depth and flow of water and, to a lesser extent, acidity or alkalinity of the water and substrate, and depth and texture of the substrate, are important factors in this community. Maximum depth of water for wool-grass is approximately 20′ (50 cm) (Kadlec, 1958) and for cattail, approximately 24′ (60 cm) (Grace & Wetzel, 1981). Higher alkalinity increases organic decomposition and nutrient availability. A slow water-flow allows accumulation of mineral and organic soil favoring growth of the rhizome and roots of these and other species associated with this community.

Graminoid Marsh Community

This community is associated with streams and ponds, becoming established in moist or saturated soil which may be flooded at times but is usually above water level. Marshes influenced by an acidic environment occur more often on the Berkshire Plateau in poorly-drained areas or in silted-up, abandoned beaver meadows. More alkaline marshes are found in the Valley Region along streams and in beaver meadows.

The vegetation is dominated by a variety of grasses, among the most typical being Canada bluejoint (*Calamagrostis canadensis*), rattlesnake grass, fowl mannagrass, northern mannagrass (*Glyceria borealis*), reed canary grass (*Phalaris arundinacea*) and rice cutgrass (*Leersia oryzoides*). Sedges are an important part of this community, with tussock sedge the most common. Many other sedges are found, with Hayden's sedge (*Carex haydenii*) a rarity, and foxtail sedge (*Carex vulpinoidea*), lakeside sedge, long-culmed sedge (*Carex folliculata*), awl-fruited sedge (*Carex stipata*) and northern bullsedge (*Scirpus atrocinctus*) fairly common as is softstem rush (*Juncus effusus*). Blue flag is common, as it is in many wetland types. Sweet flag (*Acorus calamus*) and knotted rush (*Juncus nodosus*) occur in more alkaline situations. Smaller graminoids exist between clumps of larger ones, such as spikesedges (*Eleocharis* spp.), as do many small forbs, namely St. John'sworts (*Hypericum* spp.), hedge-hyssop (*Gratiola neglecta*), swamp candles, marsh skullcap (*Scutellaria galericulata*), marsh speedwell (*Veronica scutellata*), ditch stonecrop (*Penthorum sedoides*), monkeyflower (*Mimulus ringens*), marsh bellflower (*Campanula aparinoides*), water-purslane (*Ludwigia palustris*) and tufted loosestrife (*Lysimachia thyrsiflora*), which is found in more alkaline marshes.

A sedge meadow is dominated by tussock sedge but will contain many of the same grasses noted above.

Low topography allows accumulation of organic and mineral soil. Water level is an important factor in this habitat, as the soil is saturated but above water level. Greater diversity of species is usually seen in more alkaline environments. These marshes may succeed to shrub swamps (Nichols, 1916; McVaugh, 1958) but fluctuating water levels or other disturbances such as grazing may retard vegetational change.

7. Bogs

Bog or Level Peatland

The term "bog"commonly refers to an acidic wetland distinguished by a floating mat of sphagnum moss or peat and ericaceous shrubs growing out from the margins of a pond. There are many types of these peatlands, a more accurate term, which range from peatlands that derive their water and nutrients solely from the atmosphere to those that receive both ground and surface water (Johnson, 1985). Berkshire County bogs are all level peatlands confined to basins with very little inflow of water. They are oligotrophic (low nutrient) peatlands, which mainly occur on the cool Berkshire plateau, which has a poorly-drained topography with acidic bedrock, at average elevations of 1700′ (520 m). A few bogs occur in the valley region, one particularly occurs in the sandplain over clay deposits at elevation 698′ (213 m) (Oltsch, 1974).

The vegetation may occur in a regular more or less concentric series of zones radiating outward from a central pond. Characteristically, the edge of the open water or the inner edge of the sphagnum mat is dominated by leatherleaf (*Chamaedaphne calyculata*) and water-willow (*Decodon verticillata*), whose intertwining roots and stems form a base for the sphagnum mat. Next is the open sphagnum mat, the living upper layer built on peat formed over hunderds of years from dead moss and stems of other plants. On the mat occur the many small herbaceous plants and tiny evergreen cranberry vines (*Vaccinium* spp.). Bog sedges (*Carex limosa, C. paupercula, C. canescens, C. lasiocarpa*) grow along with cotton-grasses (*Eriophorum virginicum, E. tenellum*), white beaksedge (*Rhyncospora alba*) and very rarely pod-grass (*Scheuchzeria palustris*). In this nutrient-poor environment, plants resort to carnivory for added nutrients. Clusters of insect-eating pitcherplant (*Sarracenia purpurea*) and small sundews (*Drosera rotundifolia*) occur in the sphagnum, while the bladderworts (*Utricularia cornuta, U. gibba*) prefer bare patches or borders of small pools. The cranberries (*Vacinnium macrocarpon, V. oxycoccos*) climb over the mounds of sphagnum. Several orchids add their color: rose pogonia (*Pogonia ophioglossoides*), grass pink (*Calopogon tuberosus*) and less commonly, white fringed orchis (*Platanthera blephariglottis*). Black spruce and larch occur as small trees on hummocks on the mat, but become denser and taller at the outer bog edge. The bog heaths, leatherleaf, bog rosemary (*Andromeda glaucophylla*), bog laurel (*Kalmia polifolia*) and Labrador tea (*Ledum groenlandicum*) may form a large shrub zone at the outer edge of the bog along with sweet gale (*Myrica gale*). Wild calla (*Calla palustris*) is common throughout the sphagnum while three-leaved false Solomon's seal is less so. Peatland vegetation is unique and these species are rarely found in any other community.

Several factors aid in the creation and maintenance of peatlands. Due to little inflow or outflow of water, organic acids build up in water pooled in acidic rock basins . Acidic conditions prevent much decomposition of plant material leading to a build-up of a peat layer. Sphagnum moss actively acidifies water (Andrus, 1980), producing conditions to which only a few plant species have become adapted. Basin topography creates a concentration of down-flowing cold air, and studies of a Massachusetts bog at 1772′ elevation (542 m) showed that the bog mat had approximately one third the number of frost-free days as surrounding forest (Moizuk & Livingston, 1966). The nutrient-poor, unstable, cold environment restricts deciduous tree establishment, and allows the persistence of species of more northerly distribution.

8. Calcareous Fens

Fens differ from bogs or oligotrophic peatlands in several ways. The water supply for fens comes primarily from groundwater that is quite alkaline, having flowed through calacareous bedrock. This groundwater is continually seeping through the fen, moving slowly toward a small stream (Schwintzer, 1978). High levels of minerals are continually supplied to plants. Surface water was found to range from circumneutral to alkaline (pH 6.0-8.1) by Motzkin's (1994) studies of western New England and eastern New York fens. There may be a floating mat, but sedges will be the common and dominant plants, not sphagnum. There are occasional patches of sphagnum moss, and bog vegetation, which probably occur when calcareous groundwater is not reaching that particular site. The various types of calcareous fens include forested fens, shrub fens, graminoid fens and calcareous seeps, and they can occur intermixed as a mosaic of different fen communities. Layers of peat may be present. The species composition is rich and diverse, containing many plant species that are rare in the state, and calcareous fen communities in general are uncommon in the northeast. A preponderance of calcareous fen communities in Massachusetts are found in Berkshire County because of its extensive calcareous bedrock formations, primarily Stockbridge and Walloomsac Formations.

The number of northern species found in fens may be due to cold ground water and to the topographical position of many fens that are found at the base of slopes or in basins where cold air is trapped. They may also serve as refugia since they are less apt to be developed.

Forested Fen Community

This calcareous peatland community is found more frequently in the southern half of the county, and usually is associated with other fen types. Forested fens tend to occur on wide, flat margins along small streams where calcareous groundwater is surfacing. The Schenob Brook Drainage Basin, a designated Area of Critical Environmental Concern (ACEC), has excellent examples of this and the following Shrub Fen and Sloping Graminoid Fen communities. Forested Fens around Kampoosa Fen in Stockbridge are also excellent examples.

The species composition of the tree layer shows considerable variation from site to site, there being both hardwood and conifer elements that are important. Black ash (*Fraxinus nigra*) is always present and commonly red maple also. Bur oak (*Quercus macrocarpa*) is common south of Lenox. Larch is the dominant conifer, and in the valley region, seems to be an indicator of fen conditions, although it is also the common conifer in bogs on the Plateau. In some fens hemlock is one of the dominant trees. Northern white cedar (*Thuja occidentalis*) is very rare at present, possibly because, being a valuable timber tree, it was thoroughly harvested in earlier times. Fens found at higher elevations, such as Hinsdale, may have red spruce, balsam fir, larch and black ash as codominant trees.

In forested fens trees are scattered and the canopy is open. Dark pools alternate with hummocks. Small yellow ladyslipper (*Cypripedium calceolus* var. *parviflorum*), showy ladyslipper (*C. reginae*), white adder's mouth (*Malaxis brachypoda*), fen cuckoo flower (*Cardamine pratensis* var. *palustris*), pink pyrola (*Pyrola asarifolia*), and hemlock parsley (*Conioselinum chinense*) are uncommon herbaceous species associated with forested fens.

Shrub Fens

Closely associated with other peatland fen communities, shrub fens occur at the edge or in openings of forested fens and graminoid fens. Characteristic shrubs are hoary willow (*Salix candida*), autumn willow (*Salix serissima*), and shrubby cinquefoil which is an important indicator plant of any calcareous situation, wet or dry. The small shrub alder-leaf buckthorn (*Rhamnus alnifolia*) is typical of fen conditions, and the following shrubs are usually found associated with fens: poison sumac (*Toxicodendron vernix*), smooth gooseberry (*Ribes hirtellum*), swamp red currant (*Ribes triste*), and mountain fly honeysuckle (*Lonicera villosa*). Saplings of the forested fens occur throughout the shrub fen. Herbaceous species associated with this fen community are water avens (*Geum rivale*), Labrador bedstraw (*Galium labradoricum*), swamp thistle (*Cirsium muticum*), skunk cabbage (*Symplocarpus foetidus*), fen twayblade (*Liparis loeselii*) and many other species also found in graminoid fens.

Sloping Graminoid Fen

This peatland community, also referred to as a calcareous wet meadow, is found at the edge of shrub and forested fens where cold, mineral-rich groundwater seeps to the surface, often where glacial gravels are deposited on the side of a valley.

Sedges and some grasses form the characteristic vegetation. Common species are *Carex flava, C. interior, C. granularis, C. leptalea,* and *C. hystricina*. Rare species *Carex sterilis* and *Carex tetanica* may be locally abundant. Drooping bulrush (*Scirpus pendulus*) and fen cottongrass (*Eriophorum viridicarinatum*) are more widespread. Grasses such as spike muhly and Kalm's brome (*Bromus kalmii*) are usually found in these grassy fens. Other indicative herbaceous species are water avens, Kalm's lobelia (*Lobelia kalmii*), fringed gentian (*Gentianopsis crinita*), grass of Parnassus (*Parnassia glauca*) and fen goldenrod (*Solidago purshii*).

Many of these sloping fens are presently maintained in their open meadow conditions by mowing or grazing. What kept these meadows open in presettlement days is not known, but various factors such as grazing by wild herbivores, beaver cutting and periodic flooding or fire, natural or human-related, could retard succession to shrub and forested fens.

Lake Basin Graminoid Fen

This uncommon, distinctive fen community forms on layers of peat built up in a lake basin that was possibly carved out by glacial action. Calcareous mineral-rich groundwater seeping through calcareous rock nourished thousands of years of sedge growth, which created a sedge-dominated floating mat that grew out from the edges toward the middle of the lake. The peat layer may be 9 feet (2.7m) deep (Motzkin 1994). Cores of the peat show that this community has existed for thousands of years virtually unchanged. Bogs differ by having a floating mat dominated by sphagnum mosses and extremely acidic water.

Several concentric zones of vegetation surround the central body of water, remnant of the former lake. Forested fen forms the outer edges. Between this and the open sedge mat, a narrow shrub zone dominated by the small delicate swamp birch (*Betula pumila*) occurs. The sedge mat is dominated by *Carex aquatilis* and *C. lasiocarpa*, with twig-rush (*Cladium mariscoides*) often at the water's edge. Cattail occurs in scattered fashion throughout the sedge mat. Small bare sites on the mat are habitat for alpine cottongrass (*Eriophorum alpinum*) and a rare northern species few-flowered spike-rush (*Eleocharis pauciflora*). Another northern species, northern bur-reed (*Sparganium natans*), occupies small pools. Bog elements, such as pitcher plant (*Sarracenia purpurea*), sundews (*Drosera* spp.), bog rosemary (*Andromeda glaucophylla*), large cranberry (*Vaccinium macrocarpon*) and orchids such as grass pink (*Calopogon tuberosus*) and rose pogonia (*Pogonia ophioglossoides*) occur with sphagnum moss where conditions are more acidic.

The above description is based on Kampoosa Fen in Stockbridge, an excellent and virtually undisturbed example of a Lake Basin Graminoid Fen.

Calcareous Seep Community

This community type, usually small and lacking in peat layers, is found where calcareous water seeps to the surface on a slope, forming rivulets, or runs just below the surface, in gravelly soil. This wet soil is too unstable to support any trees beyond the sapling size, therefore perpetuating sunny openings, allowing small herbaceous species to survive. Showy ladyslipper may be shaded out in a forested fen habitat, but here it can thrive for much longer periods.

The vegetation consists mainly of sedges, grasses and herbaceous species. Grass of Parnassus is a good indicator plant. Scouring rush (*Equisetum hyemale*) and variegated horsetail (*E. variegatum*) often form dense stands. Dwarf horsetail (*E. scirpoides*) occurs on cool moist mossy surfaces. Sedges *Carex aurea, C. leptalea* and *C. granularis*, the grass spike muhly, white adder's mouth, showy ladyslipper, nodding and hooded ladies'-tresses (*Spiranthes cernua, S. romanzoffiana*), sweet coltsfoot (*Petasites frigidus*), which is very rare, and Kalm's lobelia, are characteristic species.

9. Anthropogenic or human-created habitats

Human activities on the land, such as clearing, bulldozing, ploughing, herbiciding and mowing, continue to create disturbed areas that are quickly colonized by pioneer species, chiefly introduced weeds. No real natural communities arise in these fast-changing habitats, but they are important to the vegetation of the county since they provide sites for the initial colonization of new weedy species. Roadsides and railroad rights-of-way can provide fascinating places for the botanist, but alien species bode ill for our native plant communities, which are easily overwhelmed by non-natives.

One native species that took early advantage of the colonists' clearing was ragweed (*Ambrosia artemisiifolia*), whose pollen shows up in increased amounts in 17th century levels in core samples (Davis, 1965).

Roadsides

Roadsides provide open habitat, and modern wide-ranging transportation moves the easily dispersed seeds of adaptable plants. Earth-moving equipment, caked with soil, moves plants, roots and seeds from state to state. Spread primarily by this means, common reed (*Phragmites australis*) springs up on roadsides and quickly takes over marshes. Spotted knapweed (*Centaurea maculosa*) lines every sandy roadside. In a very few years, the tiny pink-flowered centaury (*Centaurium pulchellum*) has spread throughout the county. Road salting creates a new habitat for salt-tolerant plants. Seaside goldenrod (*Solidago sempervirens*) now grows along the Massachusetts Turnpike, and saltmarsh sand-spurry appears occasionally on roadsides. These are only a few of the many species colonizing roadsides.

Railroad Rights-of Way

Stressful habitats that are hot, dry, contain little soil, and are herbicided regularly provide habitat to few species, but those that can survive find little competition. Railroads are "highways" for the transportation of plant species from far and wide which hitch a ride and are dropped off along the way.

Dwarf snapdragon (*Chaenorrhinum minus*) appeared as a "railroad endemic" but soon moved to roadsides and driveways. Other exotic species turn up occasionally, such as soybean (*Glycine max*) and heartleaf umbrella-wort (*Mirabilis nyctaginea*), while natives from other areas, such as freshwater cordgrass (*Spartina pectinata*) and ashy sunflower (a prairie species) (*Helianthus mollis*) can be found.

Powerline Rights-of-Way

These areas are kept open, usually by herbiciding like railroad ROWs, but the soil is not disturbed, and there is not the transportation factor. Although also a haven for weeds, they do provide open sunny areas for native species that are less shade-tolerant. New Jersey tea (*Ceanothus americana*), winged sumac (*Rhus copallina*), stiff gentian (*Gentiana quinquefolia*) and even saplings of tuliptree (*Liriodendron tulipifera*) have been found thriving in this habitat. Blackberry, raspberry and blueberry shrubs grow under the lines, and butterfly species favor this sunny habitat.

Logging Clearings

These tend to become overgrown with brambles (*Rubus* spp.), mullein (*Verbascum thapsus*), fireweed (*Erechtites hieracifolia*), bristly sarsaparilla (*Aralia hispida*), and fire or pin cherry (*Prunus pensylvanica*). These are succeeded by early successional tree species such as white birch, black birch, quaking aspen, black cherry, red maple and possibly red oak, depending on closeness of seed source. Bigtooth aspen will remain as the Mesic Northern Hardwood Forest grows back.

Old Fields and Pastures

A cultivated field, soon after abandonment, is host to many weedy species, such as horseweed (*Conyza canadensis*), velvetleaf (*Abutilon theophrasti*), crabgrass (*Digitaria sanguinalis*) and many others.

Pastures, when abandoned, grow quickly into a dense cover of shrubby species, such as steeplebush and meadowsweet (*Spiraea tomentosa, S. latifolia*), brambles, goldenrods, and thorny shrubs hawthorn (*Crataegus* spp.) and unfortunately the aggressive multiflora rose (*Rosa multiflora*). White pine and other early successional tree species soon appear.

Abandoned pastures at high elevations (2300′, 701m) were quickly colonized by red spruce on a few saddles on Mt. Greylock and on the Berkshire Plateau (Egler, 1940). The dense red spruce stand along Sperry Road to Stony Ledge on Mount Greylock was once thought to be a rare example of a virgin northern forest (Bromley, 1935), although it had been pastured until the 1840s. These stands are now changing into a more mixed northern hardwood-spruce-fir forest.

III. History of the Botanical Exploration of the County

Beginnings in the Early 19th Century

Botanical exploration of the region was early advanced by the presence of Williams College in Williamstown. Founded in 1792, it attracted a group of educated men into the still rural, remote, undeveloped county, far from Boston and Harvard College. Two men dominated the scene in these early years. Amos Eaton and Chester Dewey, each in his own way, made valuable contributions to botanical knowledge, both of the region and beyond.

Eaton was born in 1776 in nearby Chatham, N.Y., and attended Williams College, graduating in 1799. He trained for a career in the law, but even while practicing law his real enthusiam seemed to be in natural history, primarily geology and botany. In 1810, while still active in legal affairs, he gave a series of lectures on botany. An unfortunate legal mixup (possibly due to more attention paid to science) resulted in his being charged with forgery and sentenced to life imprisonment in a New York City jail in 1811. According to Rodgers (1942) it was here that he met the young John Torrey, who accompanied his father, Fiscal Agent of the prison, on his visits to the prison. Eaton, already the prison librarian, began to teach the rudiments of botany to the teen-aged John Torrey. It was probably through the Torreys' intercession that Eaton was released in 1815 and granted a pardon in 1817 by Governor Clinton, a friend of the Torrey family (Robbins, 1968). John Torrey became an eminent botanist, compiling with Asa Gray a Flora of North America. Eaton and Torrey maintained a close lifelong friendship.

After his release Eaton wisely chose to follow his bent for teaching natural sciences, and studied under Dr. Benjamin Silliman of Yale for a year. He was invited to give a course of lectures at Williams College in 1817 on minerology and botany. His boundless enthusiasm and his talent for communicating it to his listeners made the courses tremendously popular among students and townspeople. Instead of dry exposition in the classroom, he taught in the field whenever possible, using fresh specimens and encouraging students to make their own collections (Ballard, 1897). The new Linnaean system of classification by numbers of flower parts made identification within the grasp of everyone, which suited Eaton who, in this young democracy, wished to make knowledge available to all. Enthusiasm ran so high, Hoffmann (1922) related, that the talk in the taverns at night was of the new plants seen.

Among those out scouring the hills for new specimens was Emory Washburn, a student (later to become Governor of Massachusetts) who found a new species of honeysuckle on a ledge "2 miles west of the College". Eaton named this plant *Lonicera hirsuta*, hairy honeysuckle, a vine which still twines around vegetation on the calcareous cliffs of its type locality. The exact location was unknown until rediscovered by Hoffmann in 1920. It again lapsed into obscurity until rediscovered 65 years later by the efforts of MNHESP and the author. Although the species is more common in New York State and around the Great Lakes, it is found only at three stations in Massachusetts, all in Berkshire County.

In order to have a printed text for Eaton's course, his students published his lecture notes in 1817 (Whipple, 1892). Eaton reissued the text in 1818 (Eaton, 1818), making it perhaps the first manual of botany for the northeast in English. It went through many editions, until superseded by Asa Gray's work. Gray is quoted by Ballard (1897) as saying that Eaton's manual was the only accessible work available to the general public at the time.

Inspired by his success, he became an itinerant botanical lecturer, speaking to classes in Lenox, Northampton, Amherst and Troy. He was the first to enroll women in the study of science (Ballard, 1897), and was teacher and supporter of Mary Lyon, founder of Mount Holyoke College, and Emma Willard of Troy.

His fame was such that Governor Clinton invited Eaton, who had been barred from New York State until his pardon, to give a course of lectures to the Legislature. As Ballard says, "Did an event like that ever occur before or since?". Eaton impressed the rich and powerful Stephen Van Rensselaer, who became his patron, and together they founded the school that became Rensselaer Polytechnic Institute (Rodgers, 1942). Eaton taught there until his death in 1842. He was in active correspondence with the important botanists of the day, including Dr. Bigelow, Mr. LeConte, John Torrey and Mr. Rafinesque, and left a legacy of botanists, geologists and interested amateurs.

Contemporaneous with Eaton, Chester Dewey was born in Sheffield in 1784 and graduated from Williams College in 1806. In 1810 he was appointed Professor of Mathematics and Natural Science. His main interests were botany and geology, and as Fernald (1942) notes, he became a "tireless student ... of the intricate and highly technical genus *Carex*, the sedges". Hoffmann lists nine sedge species that Dewey or other botanists had named from material collected in Berkshire County including *Carex formosa* Dewey, from Stockbridge; *Carex hitchcockiana* Dewey, from Mount Greylock; *Carex tuckermanii* Dewey, from Sheffield; *Carex davisii* Schweinitz & Torrey, from Sheffield; and *Carex schweinitzii* Dewey, in Williamstown.

During Eaton's stay in Williamstown, Dewey drew up a list of plants collected in the vicinity of the college, using Eaton's Manual. This manuscript (Dewey, 1818) listed a number of species not seen at present and documented many of the European weeds that had already become established. Flowering dates and habitat were included. A more extensive catalog of Berkshire County plants numbering about 800 species was compiled by Dewey for Field's history of the county in 1829 (Field, 1829). Several species that Dewey listed, but which had not been seen at Hoffmann's writing, have been rediscovered in the last 20 years. Among these are autumn coralroot (*Corallorhiza odontorhiza*), bitter milkwort (*Polygala polygama*), and Michaux's sedge (*Carex michauxiana*). Dewey's "Report on the Herbaceous Plants of Massachusetts", commissioned by the Governor as a companion to Emerson's report on trees and shrubs, mentioned some Berkshire plants (Dewey, 1840). Dewey's priceless collection of sedge specimens was given to Williams College, but unaccountably ended up in an attic at Bates College in Maine, and was nearly thrown into the trash to be burned. By chance, Fernald (1942) heard about this potential disaster and managed to save the specimens. Fernald's detective work revealed a series of mixups and mishaps that led to this event. Dewey is remembered as a scholarly botanist while Eaton's legacy is probably the hundreds of students inspired by his teaching, many of whom carried the spirit of inquiry westward into new lands.

Though Dewey continued to work on the sedges, active botanical exploration of the county slowed. John Torrey, William Oakes and William Boott, botanists from Boston and New York, made occasional collecting trips. J. W. Robbins collected *Potamogeton alpinus*, a northern species, in Richmond in 1964, but this species has never been reconfirmed in the county.

The interest in natural history that Eaton kindled at Williams College persisted throughout the 19th century. Influenced as a boy by Eaton's lectures, Albert Hopkins built the first Astronomical Observatory which is still extant (Ballard, 1897). A Lyceum of Natural History was established that housed the collections and involved the students. Unfortunately, some of these collections were destroyed by fire in 1855 (Gustafson, 1937). A few of Eaton's specimens remain at RPI, but most are at Yale University's Osborn Botanical Laboratory (Merrill, 1946) including the type specimen of *Lonicera hirsuta* Eat. Paul Chadbourne, President of Williams, was an enthusiastic amateur botanist and drew up a natural history catalog for Williamstown (1858). John Bascom, in his paper on the appearance and disappearance of plants (1899) discussed the increase of European weeds and the retreat of some natives, notably the Orchid Family, and made an impassioned plea for the protection of nature.

Early 20th Century Botanists

A new era opened in 1877 with the first visit of Judge J. R. Churchill, an indefatigable, knowledgeable amateur botanist from the Boston area. For fifty years he returned to the county, collecting extensively from Sheffield, Mount Greylock and Williamstown, where the "Sand Springs.....most happy hunting ground for the botanist" (Churchill, 1899) was one of his favorite haunts. He discovered several important species, such as tiny cow-lily (*Nuphar pumila*) and black cohosh (*Cimicifuga racemosa*) in Sheffield.

With the founding of the New England Botanical Club (NEBC) in 1895, interest was renewed in New England botany. Fernald collected in the county on several occasions, and in 1913 located a serpentine outcrop in the Town of Florida. The members organized into efficient teams to cover the county (Fernald, Bean & Knowlton, 1919; Knowlton, 1919). Their collections formed the basis of the NEBC Herbarium coverage of the region.

A. K. Harrison frequented the Mount Greylock region, as did A. L. Andrews (1900, 1901). Hoffmann acknowledged the contributions of F. Walters who botanized in the southern towns from 1913 to 1917. C. A. Weatherby found climbing fern (*Lygodium palmatum*) in Monterey, but it has not been seen since.

Ralph Hoffmann, compiler of the most recent Berkshire county flora (1922), was born in 1870 in Stockbridge, where his father was headmaster of a private school. He was known first as an accomplished ornithologist, publishing a paper on Berkshire birds in 1900, and subseqeuently, field guides and books on birds of a wider area. His interest in botany surfaced in 1899 (Hoffmann, 1899), and in 1904, he documented the discovery of autumn willow (*Salix serissima)* and Frank's lovegrass *(Eragrostis frankii*) in the county (Hoffmann, 1904). Thirty years of collecting laid the foundation for his flora, which is a thorough, accurate work, providing much information on habitat and distribution. His professional career was in teaching, although later in life he became Director of the Santa Barbara Museum of Natural History (Swanton, 1981). It was in 1932, while collecting plants on a coastal cliff in California, that he fell to his death.

Grace Greylock Niles, a knowledgeable, somewhat eccentric botanist, explored the northern Berkshire County/southern Vermont area for orchids and ferns and described her finds in many delightful articles and books (Niles, 1904, 1905, 1913). She usually disguised the locations under some obscure Native American place name, or invented her own. Her book ,"*Bog-trotting for Orchids*" inspired at least one orchidologist to pursue the elusive, mysterious species (Case, 1987).

Late Twentieth Century Renewal

From 1925 to 1970, there were few active botanists. A. K. Harrison was still working in the 1940s. In 1957 Hermann (1958) discovered the alien eyebright (*Euphrasia micrantha*) on the summit of Mount Greylock. Harrison had collected it in 1944, but it had been misidentified at the time. A few specimens in Becket were collected by G. N. and F. F. Jones in the 1940s. An active local botanist, S. Waldo Bailey, was for many years the Warden of Bartholomew's Cobble, a unique marble formation with a diverse flora of rare ferns (Weatherby, 1947; Arnold & Bailey, 1965). Morgan Bulkeley continues to be a fine source for locations of interesting species and areas, and has collected information on all champion trees of the county (Bulkeley, 1969).

During this period, botany ceased to be a popular avocation. Technical science became more important and it was thought that every plant had been discovered. Ecology, however, was developing into an important field, stressing the interactions of all organisms. Egler (1936, 1940) conducted a detailed ecological study of the Berkshire Plateau and its development after farm abandonment and timbering.

The belated realization that many of our natural habitats seemed to be heading for extinction spurred a renaissance in botanical exploration. An indication of renewed interest was C. Barre Hellquist's study of the relationships of pH and other dissolved substances of lakes and ponds of New England to the distribution of pondweeds (*Potamogeton* spp.) (Hellquist, 1975; 1980). He studied particularly the highly alkaline lakes of Berkshire County, documenting the presence of rare species Hill's pondweed (*Potamogeton hillii*), Fries' pondweed (*P. friesii*) and Vasey's pondweed (*P. vaseyi*).

The passage of the Endangered Species Act by the U. S. Congress in 1973, and the decision by The Nature Conservancy to focus on preserving habitat for rare species led to the inception of the Natural Heritage Program, now Massachusett Natural Heritage and Endangered Species Program (MNHESP), as well as similar organizations in most other states. The New England Botanical Club, in cooperation with the U.S. Fish and Wildlife Service, produced a first overview on what was known of uncommon and rare plant species. This report by Coddington and Field (1978) laid the groundwork for a plan of intense fieldwork to rediscover rare species collected by earlier botanists and to explore and document rare plant communities. All data were and continue to be placed into a database that is kept by MNHESP.

Skilled botanists such as Bruce A. Sorrie, then MNHESP Botanist, and Nature Conservancy botanists Caren Caljouw and Tom R. Rawinski expanded vegetation data on the county tremendously. They discovered several species in Berkshire County that were new to the state, such as smooth woodsia (*Woodsia glabella*), creeping sedge (*Carex chordorrhiza*) and chestnut sedge (*Carex castanea*).

Many others have contributed to botanical knowledge of the county. Jerry Jenkins, graduate of Williams College, has done extensive field work in northern Berkshire county and southern Vermont, rediscovering broad waterleaf (*Hydrophyllum canadense*) in the county. His imaginative and inspired teaching continues to lead many others into the field. Charles Quinlan is carrying on the tradition of field work in the under-explored Berkshire Plateau region. He had discovered mountain cranberry (*Vaccinium vitis-idaea*), the only site in the state, on the slopes of Mount Greylock.

This renewal of interest, it is hoped, will maintain its momentum. Vegetation is dynamic, and there always will be new discoveries for the field botanist. If the climate is changing, good documentation will be needed. All larger works are based on local field studies. Having walked in the footsteps of Eaton, Dewey and others, we set the prints to be followed by new workers who will find the fertile fields of Berkshire County a renewing source of wonder.

IV. CATALOG OF VASCULAR PLANTS OF BERKSHIRE COUNTY

Introduction

The catalog of vascular plants includes all species growing without cultivation in Berkshire County, including both native and alien or introduced species. Information on the native or alien status of a species is drawn chiefly from Fernald (1950) and Mitchell (1986).

The classification system used in this catalog follows that of Cronquist (1981) for Angiosperms, and Tryon and Tryon (1982) for Pteridophytes. Nomenclature follows that of Kartesz (1994) in general, but not in every case, and synonymy is provided where necessary. All taxa listed are documented by specimens, except in a very few cases, which are noted. Specimens for a majority of the taxa have been collected by the author and voucher specimens have been deposited at the Hodgdon Herbarium, University of New Hampshire (NHA). In cases where no modern specimen (since 1970) has been collected and/or the specimen is located in an herbarium other then NHA, this is noted. Certain very rare species or species uncommon in the county are documented by photographs which have been mounted and deposited in the Hodgdon Herbarium.

Information on habitat, frequency and abundance, and distribution within the county are primarily based on the field work conducted for this study, with additional information drawn from Hoffmann (1922), herbarium specimens at the New England Botanical Club (NEBC) now merged into the Gray Herbarium, University of Massachusetts (MASS), Gray Herbarium (GH), North Adams State College (NASC) and from notes of other contemporary botanists who have worked in Berkshire County, primarily those of Bruce A. Sorrie, former State Botanist. It should be stressed that this information applies only to Berkshire county, since a species may occupy different habitats, or be more or less common, in different regions.

To denote frequency, the words "common", "frequent", "occasional", and "uncommon" are used. These apply to the distribution of a species within its typical habitat, recognizing also that the habitat itself may be common or uncommon. For instance, a species that is described as "common in rich mesic forest" appears in practically all of the suitable habitat. "frequent" means appearing in most of the suitable habitat;"occasional" in less than half of the suitable habitat and occurring in five or fewer towns which are then listed, or in more towns but in very few locations; "uncommon" means represented by only a few stations anywhere in the county. The term "rare" is reserved strictly for those species officially listed as Endangered, Threatened and Special Concern Species in Massachusetts (Natural Heritage, 1994). Comments on the status of the species, whether it is declining or spreading, and possible causes of this condition are made when appropriate.

Flowering dates cited are based on field observations and from herbarium specimens. Where this information was not available, data from Seymour (1982) is used. Flowering or fruiting dates are indicated by month and week, i.e. 4(2)-5(1), meaning the second week of April to the first week of May. There can be, of course, a good deal of variability in these dates from year to year, depending on weather variations.

An English name is given whenever there is a recognized name in common usage. These names are drawn largely from Fernald (1950), Mitchell (1986), and Seymour (1982).

The statement that a species was "first discovered in the state" by a botanist refers to a species new to the state that was discovered first in Berkshire County.

Key to Symbols and Abbreviations

Taxa listed by the Massachusetts Natural Heritage and Endangered Species Program (Natural Heritage, 1994) as rare are designated by the following symbols which are placed immediately after the scientific name and author. Designations may change as more information on abundance and distribution is gathered.

E = Endangered
T = Threatened
SC = Special Concern
H = Historical, designates rare species documented from Berkshire county but for which no current stations are known in the county and hope for their rediscovery is slight.

* = denotes species which are not native to the county. They may be native to another portion of the state or North America, or be introduced from other continents.

CATALOG OF VASCULAR PLANTS

PTERIDOPHYTA (FERNS AND FERN ALLIES)

OPHIOGLOSSACEAE

BOTRYCHIUM

B. dissectum Spreng. Cut-leaf Grape Fern
Frequent. Moist pastures, open woods.

B. lanceolatum (Gmel.) Angstr. var. **angustisegmentum** Pease & Moore Lance-leaf Grape Fern

B. matricariifolium (A. Br. ex Dowell) A. Br. ex Koch Matricary Grape Fern
Occasional. Rich decidous woods. Hoffmann and specimens in NEBC indicate it was formerly more frequent. Adams, Cheshire, Lenox, North Adams, Peru, Pittsfield, Sandisfield, Sheffield.

B. multifidum (Gmel.) Rupr Leathery Grape Fern
(Including var. **intermedium** (D. C. Eaton) Farw.)
Uncommon. Old pastures, meadows. Becket, Cheshire, North Adams, Tyringham.

B. oneidense (Gilb.) House Blunt-lobed Grape Fern
Uncommon. Rich woods, open deciduous woods. Cheshire, Hancock, Lenox, Williamstown, Mt. Washington.

B. simplex E. Hitchc. var. **tenebrosum** (A. A. Eaton) Clausen Dwarf Grape Fern
Uncommon. Rich woods. Stockbridge, New Ashford.

B. virginianum (L.) Sw Rattlesnake Fern
Common. Rich, moist woods.

OPHIOGLOSSUM

O. pusillum Raf. **T** Adder's Tongue Fern
(**O. vulgatum** L.)
Rare. Hoffmann (1922) lists it as frequent. Mucky meadows, open pasture, circumneutral or calcareous soil. Great Barrington, Lanesboro, Lenox, Peru, Sheffield, Stockbridge. Plate 1A.

OSMUNDACEAE

OSMUNDA

O. cinnamomea L. Cinnamon Fern
Common. In wet or moist woods, swamps and borders of marshes.

O. claytoniana L. Interrupted Fern
Common. Moist woods.

O. regalis L. var. **spectabilis** (Willd.) Gray Royal Fern
Common. Swampy woods, shrub swamps, borders of ponds.

SCHIZAEACEAE

LYGODIUM

L. palmatum L. **H** Climbing Fern
Rare and historical. Damp, acidic soil. Monterey: <u>Townsend</u> & <u>Weatherby</u> 1925 (GH). Specimen label notes "a single plant, said to be the relic of a once considerable colony, in damp ground in open pine woods."

HYMENOPHYLLACEAE

TRICHOMANES

***T. intricatum** Farrar Filmy Fern
Uncommon. Gametophytes only, in moist, sheltered crevices. Bartholomew's Cobble, Sheffield: <u>Sorrie</u> <u>1810</u> 1983 (ISC). Campbell Falls, Sandisfield.

PTERIDACEAE

ADIANTUM

A. pedatum L. Maidenhair Fern
Common. Moist rich woods.

CRYPTOGRAMMA

C. stelleri (Gmel.) Prantl **T** Slender Rock-brake
Rare. Shaded dripping calcareous cliffs. Cheshire, West Stockbridge, Williamstown. Plate 1B.

PELLAEA

P. atropurpurea (L.) Link Purple Cliff-brake
Uncommon. Dry open or lightly shaded calcareous ledges. Not north of Lenox, occasional in Sheffield, Egremont, Lee, New Marlboro.

DENNSTAEDTIACEAE

DENNSTAEDTIA

D. punctilobula (Michx.) Moore Hay-scented Fern
Common. Open woods and clearings, acidic soil. Fragrant when warmed by sun.

PTERIDIUM

P. aquilinum (L.) Kuhn var. **latiusculum** (Desv.) Underwood Bracken
Common. Dry sterile fields, clearings and open woods.

THELYPTERIDACEAE

PHEGOPTERIS

P. connectilis (Michx.) Watt Narrow Beech Fern
(**Thelypteris phegopteris** (L.) Slosson)
Common. Moist banks, rocks near streams.

P. hexagonoptera (Michx.) Fee Broad Beech Fern
(**Thelypteris hexagonoptera** (Michx.) Weatherby)
Frequent. Moist, open woods, often oak woods sometimes somewhat rich.

P. connectilis X **P. hexagonoptera**
Uncommon. New Ashford, Williamstown. Lenox: Hill et al. 1919 (NEBC).

THELYPTERIS

T. noveboracensis (L.) Nieuwl. New York Fern
Common. Open woods, clearings.

T. palustris Schott Marsh Fern
Common. Open wet meadows, marshes, swampy woods.

T. simulata (Davenp.) Nieuwl. Massachusetts Fern
Uncommon. Acidic marshes and swampy woods near stream, occasionally in fens. Becket, Otis.

DRYOPTERIDACEAE

ATHYRIUM

A. filix-femina (L.) Roth ssp. **angustum** (Willd.) Clausen Northern Lady Fern
Common. Open moist woods and clearings.

CYSTOPTERIS

C. bulbifera (L.) Bernh. Bulblet Fern
Frequent. On or near moist shaded calcareous rock. A good indicator of calcareous conditions.

C. tenuis (Michx.) Desv. Fragile Fern
(**C. fragilis** (L.) Bernh. var. **mackayi** Laws.)
Frequent Usually on or near shaded calcareous rock outcrops, occasionally on on mortar, as on east face of Hoosac Tunnel.

DEPARIA

D. acrostichoides (Sw.) M. Kato Silvery Spleenwort
(**Athyrium thelypterioides** (Michx.) Desv.)
Frequent. Circumneutral or rich moist soil.

DIPLAZIUM

D. pycnocarpon (Spreng.) Broun Narrow-leaved Spleenwort Glade Fern
Occasional. In moist rich calcareous soil in sheltered locations such as the bases of ledges, or ravines. Adams, Cheshire, Florida, Lenox, New Ashford, Pittsfield, Williamstown.

DRYOPTERIS

D. Xbenedictii (Farw.) Wherry Benedict's Wood Fern
(**D. carthusiana** X **D. clintoniana**)
Uncommon. Wet woods and swamps. Cheshire, Lenox. New Marlboro: Hoffmann 1912 (NEBC).

D. Xbootii (Tuckerm.) Underw. Boott's Wood Fern
(**D. cristata** X **D. intermedia**)
Frequent. Swampy woods, moist banks.

D. Xburgessii Boivin Burgess' Shield Fern
(**D. clintoniana** X **D. marginalis**)
Uncommon. Wet woods. Lenox: Winslow 1925 (NEBC).

D. campyloptera (Kunze) Clarkson Mountain Wood Fern
Occasional. Cool moist woods, at higher elevations above 761m (2500′) in spruce-fir and spruce-fir-hardwood forest.

D. carthusiana (Vill.) Fuchs Spinulose Wood Fern
(**D. spinulosa** (Mull.) Watt.
Common. Forested swamps, cool damp rocky woods.

D. clintoniana (D. Eat. ex Gray) Dowell Clinton's Shield Fern
Occasional. Swampy woods in central valley region. Cheshire, Great Barrington, Lanesboro, Lenox, Pittsfield.

D. clintoniana X **D. cristata**
Uncommon. Wooded swamps. New Marlboro, North Adams, Sheffield: Churchill 1920 (NEBC).

D. cristata (L.) Gray Crested Wood Fern
Common. In wooded swamps.

D. Xdowellii (Farw.) Wherry Dowell's Wood Fern
(**D. clintoniana** X **D. intermedia**)
Uncommon. Swampy woods. Lenox, Stockbridge. Lanesboro: Knowlton 1915 (NEBC).

D. goldiana (Hook. ex Goldie) Gray Goldie's Fern
Frequent. Rich calcareous soil on slopes below ledges. Often with **Diplazium macrocarpon**.

D. goldiana X **D. intermedia**
Uncommon. Swampy woods. Monterey: Hoffmann 1920 (NEBC).

D. intermedia (Muhl.) ex Willd.) Gray Evergreen Wood Fern
Common. Moist rocky woods.

D. marginalis (L.) Gray Marginal Wood Fern
Common. Dry open rocky sites, shaded laedges or moist humus in woods.

D. Xneowherryi W. H. Wagner Wood Fern
(**D. goldiana** X **D. marginalis**)
Uncommon. Cheshire: Winslow 1912 (NEBC).

D. Xslossonae (Hahne) Wherry Slosson's Wood Fern
(**D. cristata** X **D. marginalis**)
Uncommon. Lenox: Winslow1912 (NEBC).

D. Xtriploidea Wherry Triploid Wood Fern
(**D. carthusiana** X **D. intermedia**)
Occasional. Rocky woods.

D. Xuliginosa (A. Br. ex Dowell) Druce Swamp Wood Fern
(**D. carthusiana** X **D. cristata**)
Uncommon. Shrub or forested swamps. Lenox.

GYMNOCARPIUM

G. disjunctum (Rupr.) Ching Oak Fern
(**G. dryopteris** (L.) Newman var. **disjunctum** (Rupr.) Ching)
Frequent. In cool moist woods, rich soil along streams.

MATTEUCCIA

M. struthiopteris (L.) Todaro Ostrich Fern
Common. Abundant on floodplains, alluvial soil pockets along streams.

ONOCLEA

O. sensilbilis L. Sensitive Fern
Common. Abundant in any wet area, open fields, marshes, wooded swamps.

POLYSTICHUM

P. acrostichoides (Michx.) Schott Christmas Fern
Common. Abundant in most forest habitats except the most acidic. Pinnae of some plants may exhibit a more incised form.

P. braunii (Spenner) Fee **E** Braun's Holly Fern
(Including var. **purshii** Fern.)
Rare. Edge of brooks in cold deep ravines in wet calcareous soil, on Mt. Greylock at approx. 670m (2198′). Adams, New Ashford. There are 7-8 plants at existing stations, but little regeneration seen. Has not been discovered at other stations reported by Hoffmann: Pittsfield, Florida, North Adams, Becket: Jones Becket (NY). Plate 1C.

P. Xpotteri Barrington Potter's Holly Fern
(**P. acrostichoides** X **P. braunii**)
Uncommon. Near brook in shallow ravine, Mt. Greylock, at 640m (2100′). May be occupying an intermediate habitat, less wet and calcareous, and near **P. acrostichoides** on top of bank. New Ashford. (see Barrington, 1986). First discovered in the state by the author in 1986.

WOODSIA

W. glabella R. Br. **E** Smooth Woodsia
Rare. High shaded ledge, with pockets of calcareous rock. Cheshire: Sorrie & PBW 2415 1984 (NEBC). First discovered in the state by B. A. Sorrie in 1984. Plate 1D.

W. ilvensis (L.) R. Br. Rusty Woodsia
Occasional. In crevices of exposed rock outcrops on ridge summits. Mt. Washington, Stockbridge, West Stockbridge.

W. obtusa (Spreng.) Torr. Blunt-lobed Woodsia
Occasional. On open or lightly shaded calcareous rock outcrops. Adams, Lanesboro, Pittsfield, Sheffield, Stockbridge, Williamstown.

ASPLENIACEAE

ASPLENIUM

A. Xebenoides R. R. Scott Scott's Spleenwort
(**A. platyneuron** X **A. rhizophyllum**)
Uncommon. Known only from one location: Bartholomew's Cobble, Sheffield. In shaded crevices and soil pockets of marble boulders, close to parent species. Plate 1E.

A. montanum Willd. **E** Mountain Spleenwort
Rare. Crevices in shaded quartzite ledges. Cheshire, Great Barrington and Sheffield. Plate 1F.

A. platyneuron (L.) BSP. Ebony Spleenwort
Frequent. On or near calcareous rock outcrops, in open pastures and rocky open woods.

A. rhizophyllum L. Walking Fern
(**Camptosorus rhizophyllum** (L.) Link)
Occasional to frequent. On shaded moist, moss-covered marble boulders or ledges. Large patches sometimes seen.

A. ruta-muraria L. **T** Wall-rue Spleenwort
Rare. In crevices in open sunny calcareous outcrops. Lanesboro, Sheffield, Stockbridge.

A. trichomanes L. Maidenhair Spleenwort
Frequent. Shaded calcareous and serpentine outcrops.

BLECHNACEAE

WOODWARDIA

W. areolata (L.) Moore Netted Chain Fern
In marshy ground. Mt. Washington: Moldenke 16959 1939 (NEBC). Represented by a photograph of a specimen sheet. It is thought that this may have been collected from plants introduced to Berkshire County. Not listed by Hoffmann (1922).

W. virginica (L.) Sm. Virginia Chain Fern
Stockbridge 1833 (NEBC). Not listed by Hoffmann.

POLYPODIACEAE

POLYPODIUM

P. appalachianum Haufler & Windham Rock Polypody
Common. On acidic rock boulders and crevices in cliff faces in open shade. Occasionally seen on calcareous outcrops. (see Haufler & Windham, 1991)

MARSILEACEAE

MARSILEA

***M. quadrifolia** L. Water-clover
Uncommon. In quiet shallow water, rooted in deep mud. Egremont. Introduced from Europe.

EQUISETACEAE

EQUISETUM

E. arvense L. Field Horsetail
Common. Waste areas, stream banks, fields, roadsides.

E. fluviatile L. River Horsetail
Common. Marshes, wet meadows and shallow water.

E. hyemale L. var. **affine** (Engelm.) A. Eat. Scouring Rush
Common. In large patches in open woods where seepage occurs, moist banks. Evergreen.

E. pratense Ehrh. Meadow Horsetail
Uncommon. Sandy loam at edge of woods, moist rich soil in open woods. Egremont, New Marlboro.

E. scirpoides Michx. **SC** Dwarf Scouring Rush
Rare. Calcareous seepage areas on shaded cool banks, drier hummocks in fens. Adams, Cheshire, Lanesboro, Lee, Lenox, New Marlboro, Stockbridge and Williamstown.

E. sylvaticum L. Wood Horsetail
Frequent. Moist or swampy woods, more often in circumneutral soil.

E. variegatum Schleich. ex Weber & Mohr
Variegated Horsetail
Frequent. Moist open seepage slopes, roadsides, floors of marble quarries, moist open disturbed areas. Often in large patches.

LYCOPODIACEAE

DIPHASIASTRUM

D. digitatum (A. Br.) Holub Ground Cedar
(**Lycopodium digitatum** A. Br.)
Common. Open acidic woods, clearings and old fields.

D. Xhabereri House
(**D. digitatum** X **D. tristachyum**)
Uncommon. Open acidic woods, clearings. Usually found with both parents. Cheshire, Monterey, Williamstown.

D. tristachyum (Pursh) Holub Slender Ground Cedar
(**Lycopodium tristachyum** Pursh)
Occasional. Clearings on open ridgetops. Cheshire, Hancock, Pittsfield, Sheffield, Williamstown.

HUPERZIA

H. appalachiana Beitel & Mickel Appalchian Firmoss
Uncommon. Open rocky habitat at high elevations (870m, 2850′) on Mt. Greylock. Adams: Burnham 1909 S (Stockholm). This species has recently been separated from **H. selago**. (Beitel & Mickel, 1992). The site has not been relocated yet.

H. Xbuttersii (Abbe)
(**H. selago** X **H. lucidula**)
Uncommon. Steep cold northeast slope of Mt. Greylock at approx. 609m (2000′).

H. lucidula (Michx.) Trev. Shining Clubmoss
(**Lycopodium lucidulum** Michx.)
Common. Cool moist woods. Abundant in higher elevations in spruce-fir-hardwood forests.

H. selago (L.) C. Mart. & Schrank **H** Firmoss
(**Lycopodium selago** L.)
Rare and historical. Andrews (1901) reported finding this species growing on bank of cold stream about 870m (2000′)

"near the road to the Bellow's Pipe". The specimen, which Hoffmann (1922) stated was in the Williams College Herbarium, is now in the Muskingum College Herbarium, Muskingum College, Muskingum, Ohio.

LYCOPODIELLA

L inundata (L.) Holub Bog Clubmoss
(**Lycopodium inundatum** L.)
Frequent. On open bog mat, also in damp sand. Monterey, Otis, Sheffield, Washington, Williamstown.

LYCOPODIUM

L. annotinum L. var. **annotinum** Bristly Clubmoss
Common. Acidic, open upland woods.

L. annotinum L. var. **acrifolium** Fern. Bristly Clubmoss
Occasional. Dry open acidic woods. Lanesboro, Otis.

L. clavatum L. Staghorn Clubmoss
Common. Acidic open woods.

L. dendroideum Michx. Northern Tree Clubmoss
Common. Dry open acidic woods.

L. hickeyi W. H. Wagner, Beitel & Moran Hickey's Clubmoss
(**L. obscurum** var. **isophyllum** Hickey)
Occasional. In less acidic soil, probably more common but under-collected. Mt. Washington, Monterey.

L. obscurum L. Eastern Tree Clubmoss
Common. Open acidic woods.

SELAGINELLACEAE

SELAGINELLA

S. apoda (L.) Fern. Creeping Spikemoss
Common. Moist rich soil in pastures, wet meadows, lawns, gardens, hillsides in the central valley region.

S. rupestris (L.) Spring Rock Spikemoss
Uncommon. Exposed marble and serpentine outcrops. Florida, Sheffield: Sorrie, Woolsey and Boutard 1203 1981 (NEBC).

ISOETACEAE

ISOETES

I. muricata Durieu Quillwort
(**I. echinospora** Durieu)
Frequent. Acidic ponds, Berkshire Plateau and Taconic Range.

PINOPHYTA (GYMNOSPERMS)

TAXACEAE

TAXUS

T. canadensis Marsh. American Yew
Common. Cool woods, steep moist slopes, wet woods. More common at higher elevations.

PINACEAE

ABIES

A. balsamea (L.) Mill. Balsam Fir
Common above 792m (2600′) on Mt. Greylock, where it is the dominant tree at the summit at 1064m (3491'), and on the Berkshire Plateau in conifer swamps, bogs and cool ravines. Rarely seen in the valleys.

LARIX

***L. decidua** Mill. European Larch
Occasional. Moist soil. Monterey, Mt. Washington, Williamstown. Introduced from Europe.

L. laricina (DuRoi) Koch Tamarack, American Larch
Frequent. Forested fens in the central valley region, on calcareous seepage slopes and in sphagnum bogs on the Berkshire Plateau.

PICEA

***P. abies** (L.) Karst. Norway Spruce
Frequent. Widely planted and self-seeding. Introduced from Europe.

P. mariana (Mill.) BSP. Black Spruce
Occasional. Margins of sphagnum bogs on the Berkshire Plateau.

P. rubens Sarg. Red Spruce
Occasional. Scattered in lowland forests, becoming more common above 457m (1500′). Common on upper slopes of Mt. Greylock, where there are some old growth patches on the steep sides of the Hopper. Has seeded into abandoned pastures on Mt. Greylock and the Berkshire Plateau, where it also occurs with northern hardwoods. A few patches on the Taconic Range.

PINUS

P. resinosa Soland. ex Ait. Red Pine
Uncommon. Indigenous only on a few open rocky ridges, such as Tom Ball Mountain. Can withstand fire and may need it to provide open areas for seedlings. Great Barrington, Sheffield, Stockbridge, Alford-West Stockbridge. Widely planted.

P. rigida Mill. Pitch Pine
Occasional. Open rocky ridges, sandplain and flat summits. Common on the sandplain in Sheffield, but declining due to lack of fire. Dwarf trees seen on Mt. Everett, Sheffield, and Pine Cobble Mtn in Clarksburg and Williamstown.

P. strobus L. White Pine
Common. Sandy soil in lowlands of valleys, old fields and disturbed areas. Existing as tall isolated specimens in mature upland forest in northern half of county, more widespread in southern half.

***P. sylvestris** L. Scotch Pine
Occasional. Widely planted, self-seeding into disturbed areas. Introduced from Europe.

TSUGA

T. canadensis (L.) Carr. Eastern Hemlock
Common. In most forests. Cool ravines, steep rocky north-facing slopes and poorly drained areas in thin or poor soil.

CUPRESSACEAE

JUNIPERUS

J. communis L. Common Juniper
Frequent. Dry soil in old fields.

J. virginiana L. Red Cedar Juniper
Frequent. Dry soil in old fields. Abundant in calcareous soil and ledges at Bartholomew's Cobble, Sheffield.

THUJA

T. occidentalis L. **E** Northern White Cedar, Arbor Vitae
Rare. Although Hoffmann stated that there were no native stands in the county, several stations in fens in Sheffield are judged to be indigenous (Sorrie, 1989). It is also widely planted and self-seeds to open calcareous outcrops and marble quarries.

MAGNOLIOPSIDA (DICOTYLEDONS)

MAGNOLIACEAE

LIRIODENDRON

L. tulipifera L. Tulip Tree
Occasional. Rich, moist soil. Found only from Lenox south in central valley region. Also planted.

LAURACEAE

LINDERA

L. benzoin (L.) Blume Spicebush
Frequent. Circumneutral wooded swamps, generally in the central valley region. Fl. 4(4).

SASSAFRAS

S. albidum (Nutt.) Nees Sassafras
Frequent. South-facing slopes in open, dry woodland. Generally in the central valley region, except for an occurrence in Florida on serpentine ledges. Fl. 5(3).

ARISTOLOCHIACEAE

ASARUM

A. canadense L. var. **canadense** Wild Ginger
Common. Rich woods Fl. 4(3)-6(1). Less common on Berkshire Plateau.

A. canadense var. **acuminatum** Ashe Wild Ginger
Uncommon. Rich woods. Fl. 4(3)-6(1). Adams, N. Adams, Sheffield. Mt. Washington: Hunnewell 1924 (NEBC).

NYMPHAEACEAE

NUPHAR

N. pumila (Timm) DC. **T** Tiny Cow Lily
(**N. lutea** (L.) Sibth. & Sm. ssp. **pumila** (Timm) Beal; **N. microphylla** (Pers.) Fern.)
Rare. Quiet shallow waters of oxbow ponds along the Housatonic River. Fl. 7(3-4)-8(1). Sandisfield, Sheffield, Stockbridge. Plate 2A.

N. rubrodisca Morong
(**N. lutea** (L.) Smith ssp. **rubrodisca** (Morong) Hellquist & Wiersema)
Uncommon. Acidic ponds. Becket, Sandisfield: Hoffmann 1912.

N. variegata Engelm. ex Durand Yellow Pond-lily
(**N. lutea** (L.) Sibth. & Sm. ssp. **variegata** (Dur.) Beal)
Common. In both acidic and alkaline ponds. Fl. 6-8.

NYMPHAEA

N. odorata Ait. ssp. **odorata** Fragrant Water-lily
Common. In both acidic ponds in bogs and alkaline ponds. Fl. 6(2)-8.

*__N. odorata__ Ait. ssp. **tuberosa** (Paine) Wiersema & Hellquist
(**N. tuberosa** Paine) Water-lily
Uncommon. In alkaline ponds and lakes. Fl. 8. Lanesboro. Stockbridge: Hellquist 7389 (NEBC). Thought to be spreading from plants introduced from Michigan (Sorrie, 1987). First discovered in the county by C. B. Hellquist in 1973.

CABOMBACEAE

BRASENIA

B. schreberi Gmel. Water-shield
Frequent. Acidic to circumneutral ponds. Fl. 6(3)-8.

CABOMBA

***C. caroliniana** Gray Fanwort
Uncommon. In Mausert's Pond, Clarksburg. Fl. 8(3). A potentially invasive weed, adventive from south. First discovered in the county by the author in 1994.

CERATOPHYLLACEAE

CERATOPHYLLUM

C. demersum L. Coontail, Hornwort
Common, in circumneutral to alkaline ponds.

C. echinatum Gray Hornwort
Occasional. Acidic to circumneutral ponds. Otis, Stockbridge. New Marlboro. First discovered in the county by C. B. Hellquist in 1974.

RANUNCULACEAE

ACTAEA

A. Xludovicii Boivin
(**A. pachypoda X A. rubra**)

Occasional. Moist woods. Monterey, Lenox. Mt. Washington. Red fruits with thick pedicels or coral red fruits with thin pedicels.

A. pachypoda Ell. Doll's Eyes, White Baneberry
Common, in moist woods, rich woods. Fl. 5(2-4).

A. rubra (Ait.) Willd. Red Baneberry
Common. Rich woods and roadsides. Fl. 5(1-3).

ANEMONE

A. canadensis L. Canada Anemone
Occasional. Moist or alluvial meadows. Fl. 6(1) Great Barrington, Richmond, Sheffield.

A. cylindrica Gray Thimbleweed
Frequent. Dry, open, somewhat limy old fields. Fl. 7(2).

A. quinquefolia L. Wood Anemone
Frequent. Moist woodlands. Fl. 5(2).

A. virginiana L. var. **virginiana** Tall Anenome
Frequent, in edges of moist woodland, roadsides. Fl. 7(2).

A. virginiana L. var. **riparia** (Fern.) Boivin Tall Anemone
(**A. riparia** Fern.)
Occasional. Edges of moist woodland, roadsides, richer sites than the nominate. Fl. 6(1).

ANEMONELLA

A. thalictroides (L.) Spach Rue Anemone
(**Thalictrum thalictroides** (L.) Eames & Boivin)
Occasional. Rich mesic forest. Fl. 5(2). Great Barrington, Sheffield, Stockbridge.

AQUILEGIA

A. canadensis L. Columbine
Frequent. Open rocky ledges, rocky woods, frequently calcareous. Abundant on open serpentine ledge, Florida. Fl. 5(2)-5(4). Hoffmann noted two forms: f. **flaviflora** (Tenney) Britt., a yellow-flowered form, and f. **phippenii** (J. Robins.) R. Hoffm., a salmon-pink form.

A. vulgaris L. Blue Columbine
Occasional. Fl. 6-8. Mt. Washington: Hoffmann 1915 (NEBC). Introduced from Europe, a garden escape.

CALTHA

C. palustris L. Marsh Marigold
Common, and abundant in open wet meadows along brooks, shrub swamps and open swampy woods. Fl. 4(4)-6(3).

CIMICIFUGA

C. racemosa (L.) Nutt. **E** Black Cohosh
Rare. In open, rich woods. Fl. 7(2). Sheffield. Hoffmann noted it was "frequent in the western part of Sheffield" and had not been noted by Dewey (In Field, 1829). Two stations known at present, plants are mostly vegetative, possibly being shaded out by forest succession.

CLEMATIS

C. occidentalis (Hornem.) DC. **SC** Purple Clematis
(**C. verticillaris** DC.)
Rare. On open calcareous ledges. Fl. 5(3). Lanesboro, Mt. Washington, Sheffield. West Stockbridge, Williamstown.

C. virginiana L. Virgin's Bower
Common. Shrub swamps, hedgerows in moist soil. Fl. 8(1-3).

COPTIS

C. trifolia (L.) Salisb. Goldthread
(**C. groenlandica** (Oeder) Fern.)
Common. Moist, cool acidic woods. Fl. 5(1-3).

HEPATICA

H. acutiloba DC. Sharp-lobed Hepatica
(**H. nobilis** P. Mill. var. **acuta** (Pursh) Steyermark)
Frequent. Moist rich woods. Fl. 4(2-4).

H. americana (DC.) Ker. Blunt-lobed Hepatica
(**H. nobilis** P. Mill. var. **obtusa** (Pursh) Steyermark)
Frequent. Dry rich soil or more acidic soil than preceding, in open woods. Fl. 4(2-4).

HYDRASTIS

H. canadensis L. **E** Golden-seal
Rare. Rich, moist soil, base of limestone ledges. Stockbridge. First discovered in the county by B. A. Sorrie in 1988.

RANUNCULUS

R. abortivus L. Kidneyleaf Crowfoot
Common. Moist, rich woods, clearings. Fl. 4(3-4).

***R. acris** L. Common Buttercup
Common. Meadows, fields, roadsides. Fl. 6-8. Introduced from Europe.

R. allegheniensis Britt. Allegheny Crowfoot
Uncommon. Moist woods, somewhat rich. Fl. 5(3-4). Sheffield. Mt. Washington.

***R. bulbosus** L. Bulbous Crowfoot

Frequent. Dry, often calcareous soil. Fl. 6(1)-7(3). Introduced from Europe.

R. flabellaris Raf. ex Bigel Yellow Water-crowfoot
Occasional. Pools in wooded swamps and quiet water along rivers. Lenox and south.

R. hispidus Michx. var. **hispidus** Hispid Buttercup
Uncommon. Rich dry or moist open woods. Fl. 5(2). Sheffield. Plate 2B.

R. hispidus Michx. var. **caricetorum** (Greene) Duncan
(**R. septentrionalis** Poir.) Swamp Buttercup
Common. Wooded swamps, stream banks, moist floodplains. Fl. 5(1)-7(2).

R. longirostris Godr. White Water-crowfoot
Frequent. Quiet, alkaline water of ponds and slow streams. Lanesboro and south in central valley region. Fl. 6(3)-8(3). The form of plants growing in moving water approach that of **R. trichophyllus**. (See Voss, 1985). Plate 2C.

R. micranthus Nutt. ex Torrey & A. Gray **T**
Tiny-flowered Crowfoot
Rare. Open rocky oak-hickory woods on south-facing slope, in rock crevices. Fl. 5(2-4). First found in the county by the author in 1990.

R. pensylvanicus L.f. **T** Bristly Buttercup
Rare. Moist alluvial soil, in open areas or open swampy woods. Fl. 7(1). Cheshire, Lanesboro, New Marlboro, Pittsfield, Sheffield, Stockbridge, W. Stockbridge. Formerly frequent according to Hoffmann.

R. recurvatus Poir. ex Lam. Hooked Crowfoot
Common. Wet places in woods, near springs, along trails. Fl. 5(3).

***R. repens** L. var. **repens** Creeping Buttercup
Common. Moist streamsides, in ditches. Fl. 6(1)-8(1). Introduced from Europe.

***R. repens** L. var. **pleniflorus** Fern. Double Buttercup
Uncommon. Fl. 5(4). Hancock: Bean & White 1919 (NEBC). Introduced from Europe.

R. sceleratus L. var. **sceleratus** Cursed Crowfoot
Uncommon. On mud in pools between vegetation in marsh, marshy pasture. Fl. 6(1-2). Sheffield..

THALICTRUM

T. dioicum L. Early Meadow Rue
Frequent. Rich rocky soil and rock outcrops. Fl. 4(4)-5(2).

T. pubescens Pursh Tall Meadow Rue
(**T. polygamum** Muhl.)
Common. Wet meadows, wooded swamps, borders of streams in woods up to high elevations (914 m). Fl. 6(1-3).

T. revolutum DC. Waxy Meadow Rue
Uncommon. Wooded rocky hillsides. Fl. 6(1)-7(1). Sheffield, Monterey: Hoffmann 1920 (NEBC).

BERBERIDACEAE

BERBERIS

***B. thunbergii** DC. Japanese Barberry
Common and abundant.A weedy pest in old pastures and open woodlands. Fl. 5(3). Introduced from Europe.

***B. vulgaris** L. Common Barberry
Frequent. A weed in old pastures, hedgerows, roadsides. Fl. 5(4). Hoffmann (1922) noted that both these species "were becoming established". Introduced from Europe.

CAULOPHYLLUM

C. thalictroides (L.) Michx. Blue Cohosh
(Including **C. giganteum** (Farw.) Loconte & Blackwell)
Frequent. Moist rich woods, of which it is a good indicator. Fl. 4(3)-5(1).

PODOPHYLLUM

***P. peltatum** L. May Apple
Occasional. Moist rich woods. Fl. 5(2)-6(4). Florida: E. Hellquist 254 1985 (NASC). Planted and escaped. Native further south and west, although Hoffmann believed it was native here in southern part of county.

LARDIZABALACEAE

AKEBIA

*__A. Xpentaphylla__ (Mak.) Mak. Five-leaf Akebia
Uncommon. Open woods by stream. Sheffield: Sorrie et al. 2313 1983 (MASS). Introduced from Asia, a garden escape.

MENISPERMACEAE

MENISPERMUM

M. canadense L. Moonseed
Occasional. Rich alluvial woods and thickets. Fl. 6(3). Great Barrington, Sheffield.

PAPAVERACEAE

CHELIDONIUM

***C. majus** L. Greater Celandine
Common. Moist soil around buildings and waste ground. Fl. 5(2). Introduced from Europe.

SANGUINARIA

S. canadensis L. Bloodroot
Frequent. Moist alluvial banks, moist rich woods and edges. Fl. 4(2-4).

FUMARIACEAE

ADLUMIA

A. fungosa (Ait.) Greene ex BSP. **T** Climbing Fumitory
Rare. On open talus boulders, usually quartzite. Fl. 6(4)-9(4). Great Barrington. New Marlboro, Mt. Washington, Stockbridge, Williamstown. Plate 2D.

CORYDALIS

C. sempervirens (L.) Pers. Pale Corydalis
Occasional. Open rocky ridges, in crevices of bare rock. Fl. 5(3)-9(2). Great Barrington, Mt. Washington, Sheffield, West Stockbridge. Plate 2E.

DICENTRA

D. canadensis (Goldie) Walp. Squirrel Corn
Occasional. Cool rich moist woods. Fl. 4(4)-5(4). Northern half of county.

D. cucullaria (L.) Bernh. Dutchman's Breeches
Occasional. Moist rich woods. Fl. 4(4)-5(4). Most often seen in northern half of county.

CERCIDIPHYLLACEAE

CERCIDIPHYLLUM

***C. japonicum** Sieb. & Zucc. Katsura Tree
Uncommon. Disturbed moist ground. A small tree, 3 m tall, in a moist disturbed area near an old estate, with other alien species. Did not appear to be planted. Introduced from Asia. First discovered in the county by the author in 1988.

PLATANACEAE

PLATANUS

P. occidentalis L. Sycamore
Common. In alluvial soil, on banks of rivers and streams.

HAMAMELIDACEAE

HAMAMELIS

H. virginiana L. Witch-hazel
Common. Open dry or moist woods, edges of woods. Fl. 9(4)-10(3). Not common on the Berkshire Plateau (Hoffmann, 1922).

ULMACEAE

CELTIS

C. occidentalis L. Hackberry
Occasional. Alluvial woods on the higher terraces of floodplains. Great Barrington, Stockbridge. Sheffield.

ULMUS

U. americana L. American Elm
Common. Moist alluvial soil, in floodplains and along streams, in hedgerows and roadsides. Fl. 4(3). Not common on the Berkshire Plateau. Was widely planted as a street tree, now most large specimens have been killed by Dutch Elm Disease.

***U. pumila** L. Siberian Elm
Uncommon. Waste area. Pittsfield: Ahles 78511 1973 (MASS). Introduced from Asia.

U. rubra Muhl. Slippery Elm
Frequent. Alluvial soil on river banks and wet soil along streams. Fl. 4(3). Not found on the Berkshire Plateau.

***U. glabra** Hudson Wych Elm
Uncommon. In rich moist woods near settled areas.

CANNABACEAE

HUMULUS

***H. japonicus** Sieb. & Zucc. Japanese Hops
Uncommon. In open alluvial soil on bank of Housatonic R. Sheffield. Fl. 8(2-4). From eastern Asia. First discovered in the county by the author in 1992.

***H. lupulus** L. Common Hops
Occasional. Waste areas, around houses, in thickets and hedgerows. Fl. 9(3). Introduced from Europe.

MORACEAE

MORUS

***M. alba** L. White Mulberry
Uncommon. Escaped from planted specimens. Adams, Egremont, North Adams, Sheffield. Introduced from Asia.

M. rubra L. **E** Red Mulberry

Rare. Rich soil below calcareous ledges. New Marlboro, Sheffield: Sorrie 4400 1988 (MASS). Native and also occasionally planted.

URTICACEAE

BOEHMERIA

B. cylindrica (L.) Sw. False Nettle
Frequent. Marshes, edges of shrub swamps. Fl. 7(4)-9(3).

LAPORTEA

L. canadensis L. Wood Nettle
Common. Rich woods, floodplains, streambanks. Fl. 7(1)-9(3).

PARIETARIA

P. pensylvanica Muhl. ex Willd. Pellitory
Uncommon. Open or shaded calcareous ledges. Fl. 8. Great Barrington, New Marlboro, Sheffield, Stockbridge.

PILEA

P. fontana (Lunell) Rydb. Clearweed
Uncommon. Edges of marshes. Fr. 9(3)-10(1). Monterey, Stockbridge. First discovered in the county by the author in 1989.

P. pumila (L.) Gray Clearweed
Common. Wet places in woods, weed in moist ground. Fr. 9(3)-10(1).

URTICA

***U. dioica** L. ssp. **dioica** Stinging Nettle
Frequent. Floodplains, moist, rich pastures. Fl. 7-9.

U. dioica L. ssp. **gracilis** (Ait.) Selander Stinging Nettle
Common. On roadsides, hedgerows. Fl. 7(2)-9.

JUGLANDACEAE

CARYA

C. cordiformis (Wang.) Koch Bitternut Hickory
Frequent. Moist rich woods, moist woods. Fl. 5(3). In the central valley region.

C. glabra (Mill.) Sweet Pignut Hickory
(Including varieties)
Frequent. Dry woods, open ridges. Fl. 5(3-4). Great Barrington south.

C. ovalis (Wang.) Sarg. Sweet Pignut
Uncommon. Dry rich woods on south-facing slopes. Fl. 5(3). Williamstown. This species is difficult to separate from **C. glabra** unless mature fruit, leaves, bark and twigs are available. There is considerable disagreement on the **C. glabra-ovalis** complex according to Mitchell (1988), who provided a detailed description of **C. ovalis.** Manning (1950, 1973) also dicussed the two species in detail. Due to the difficulty of separating the two species, its distribution in the county may be overlooked.

C. ovata (Mill.) Koch Shagbark Hickory
Frequent. Dry rich woods, ridges, hedgerows. Fl. 5(4).

JUGLANS

J. cinerea L. Butternut
Common. Moist woods, river banks, upper edges of floodplains. Fl. 5(3).

***J. nigra** L. Black Walnut
Uncommon. Rich moist soil. Fl. 5(3). Sheffield. Native farther west, but widely planted in the county and self-seeding abundantly from trees planted along a road.

MYRICACEAE

COMPTONIA

C. peregrina (L.) Coult. Sweet-fern
(**Myrica asplenifolia** L.)
Common. Dry sterile soil in clearings,fields, open woods. Fl. 6(1).

MYRICA

M. gale L. Sweet-gale
Frequent. Edges of bogs, ponds and marshes. Fl. 5(1).

FAGACEAE

CASTANEA

C. dentata (Marsh.) Borkh. American Chestnut
Frequent. Sprouts from stumps of trees killed by chestnut blight. In dry open oak woods. Fl. 7(1-2). Occasional trees reach 12-20 cm DBH, flower and bear burs but no viable nuts seen.

FAGUS

F. grandifolia Ehrh. American Beech
Common. Dry or moist, rich or acidic woods. Fl. 5(1-2).

QUERCUS

Q. alba L. White Oak

Common. Widespread in the southern part of the county in dry woods. Occasional in the north on dry, rich, south-facing slopes. Fl. 5(3). Not on the Berkshire Plateau.

Q. bicolor Willd. Swamp White Oak
Occasional. Along streams and rivers, in swamps. Fl. 5(2-4). Egremont. Sheffield.

Q. coccinea Muenchh. Scarlet Oak
Uncommon. Dry soil in open woods. Fl. 5(1-3). New Marlboro. Sheffield, Williamstown.

Q. ilicifolia Wang. Scrub Oak
Occasional. Sand plain in Sheffield and open rocky summits of southern part of county. Fl. 5(2-4). Alford, Mt. Washington, New Marlboro, Sheffield, West Stockbridge.

Q. macrocarpa Michx. **SC** Bur Oak, Mossy-cup Oak
Rare. Forested fens, rich alluvial soil along rivers, along roads at edge of wet calcareous meadows. Fl. 6(1). Lenox and Stockbridge south in central valley region.

Q. muhlenbergii Engelm. **SC** Yellow Oak, Chinquapin Oak
Rare. On dry calcareous ridges and open slopes. Fl. 5(3). Egremont, Great Barrrington, Sheffield.

Q. prinoides Willd. Dwarf Chinquapin Oak
Uncommon. Dry sandy soil in clearings and roadsides on sand plain. Fl. 5(4). New Marlboro, Sheffield.

Q. montana Willd. Chestnut Oak
(**Q. prinus** L.)
Frequent. Rocky, upper slopes of ridges and mountains. Fl. 5(3). Williamstown, North Adams, and from Stockbridge south.

Q. rubra L. Red Oak
Common. Abundant in acidic or rich woods. In northern county, most common on south-facing slopes, and more widespread in southern county. Not at elevations over 760 m. The only oak on the Berkshire Plateau (Hoffmann, 1922) and not common there.

Q. Xsaulii Schneid. Saul's Oak
(**Q. montana** X **Q. alba**)
Uncommon. Found where both parents occur. Williamstown: Churchill 1897 (NEBC).

Q. Xschuettii Trel.
(**Q. bicolor** X **Q. macrocarpa**)
Uncommon. Wooded swamp with both parents present. Sheffield.

Q. velutina Lam. Black Oak
Frequent. Dry, rocky slopes. Fl. 5(2-4).

BETULACEAE

ALNUS

A. incana (L.) Moench ssp. **rugosa** (Du Roi) Clausen
(**A. rugosa** (Du Roi) Spreng.) Speckled Alder
Common. Shrub swamps, wet soil, ditches. Often forming pure stands. More common on the Berkshire Plateau.

A. viridis (Chaix) DC. ssp. **crispa** (Ait.) Turrill **SC**
(**A. crispa** (Ait.) Pursh) Green Alder, Mountain Alder
Rare. Edge of Deerfield River and on rocky ledges above river. Fl. 5(2). Florida.

A. serrulata (Dryand. ex Ait.) Turrill Smooth Alder
Frequent. Edges of marshes, ponds.

BETULA

B. alleghaniensis Britt. Yellow Birch
(**B. lutea** Michx.f.)
Common. Cool moist woods, forested swamps, to the summit of Mt. Greylock where it is quite shrub-like. Fl. 5(1).

B. lenta L. Sweet Birch, Black Birch
Common. Moist or dry, rocky woods. More common in cut-over or disturbed forest and at low elevations (below 600 m, 2000'). Fl. 5(3).

B. papyrifera Marsh. var. **papyrifera** Paper Birch
Common. Moist or dry woods, rocky ridges, edges of woods, hedgerows, old fields. More common where vegetation has been disturbed by agriculture, timbering or fire.

B. papyrifera Marsh var. **cordifolia** (Regel) Fern.
Heart-leaved Paper Birch
Uncommon. Upper elevations of Mt. Greylock, in spruce-fir and spruce-fir-hardwoods forest. New Ashford, Adams.

***B. pendula** Roth European Birch
Uncommon. Pittsfield: Ahles 78407 1973 (MASS). Introduced from Europe.

B. populifolia Marsh. Gray Birch
Common. Sterile soil in old fields, roadsides, open and disturbed areas. Fl. 5(1). Short-lived and soon dying as forest matures.

B. pumila L. **T** Swamp Birch
Rare. At edge of shrub and forested fens. New Marlborough, Sheffield. Stockbridge.

CARPINUS
C. caroliniana Walt. Ironwood, Hornbeam, Musclewood

Common. Moist soil in open woods, edges of woods, early successional forest, floodplains.

CORYLUS

C. americana Walt. Hazelnut
Occasional. Sandy soil at edge of open woods, clearings, roadsides. Fl. 4(2). Lenox, Sheffield.

C. cornuta Marsh. Beaked Hazelnut
Frequent. Open dry woods, clearings, thicket edges. Fl. 4(2).

OSTRYA

O. virginiana (Mill.) Koch Hop Hornbeam
Common. Rich well-drained soil, dry rich ridgetops.

PHYTOLACCACEAE

PHYTOLACCA

P. americana L. Pokeweed
Common. Moist disturbed soil, roadsides and around houses. Fl. 7(2)-9.

NYCTAGINACEAE

MIRABILIS

*__M. nyctaginea__ (Michx.) MacM. Heartleaf Umbrella-wort
Occasional. Along railroad tracks. Fl. 9(1). Cheshire, Williamstown. Native farther west.

CHENOPODIACEAE

ATRIPLEX

*__A. patula__ L. ssp. **hastata** (L.) Hall & Clements
Halberd-leaved Orach
Uncommon. Alluvial soil in floodplain in pasture. Fl. 9(2). Considered by Hoffmann to be a "fugitive species" but apparently it persists. Native to coastal areas.

CHENOPODIUM

*__C. album__ L. Lamb's-quarters, Pigweed
Common. Waste ground, roadsides, stream banks, garden weed. Fl. 8-9. Introduced from Europe.

*__C. ambrosioides__ L. Mexican-tea
Uncommon. Gravel and cobble beach along Hoosic R. Williamstown. First discovered in the county by the author in 1993. From tropical America.

*__C. botrys__ L. Jerusalem Oak
Uncommon. Waste ground. Fl. 7(3)-9). Cheshire: Hoffmann 1920 (NEBC). Introduced from Europe.

*__C. bushianum__ Aellen Pigweed
Uncommon. Waste ground. Fl. 8(2)-9. Lee: Hoffmann 1920 (NEBC). Introduced from Europe.

*__C. capitatum__ (L.) Aschers. Strawberry-blite
Uncommon. Disturbed soil. Fl. 7-8. Mt. Washington: A. Williams 1920 (NEBC). Although native to North America, stations in Berkshire County indicate it is adventive here.

C. foggii H. A. Wahl Fogg's Goosefoot
Uncommon. Shaded calcareous ledges. Fl. 8(4)-9(2). Mt. Washington. New Marlboro: Hoffmann 1920 (NEBC).

C. gigantospermum Aellen Maple-leaf Goosefoot
(**C. hybridum** L. ssp. **gigantospermum** (Aellen) Hulten)
Uncommon. Rich soil, base of calcareous ledges. Fl. 8(4). Mt. Washington, Sheffield, Stockbridge.

*__C. glaucum__ L. Oak-leaf Goosefoot
Uncommon. Waste areas, mudflats. Fl. 9(1-2) Lee, Pittsfield. Introduced from Eurasia.

*__C. standleyanum__ Aellen Goosefoot
Uncommon. Waste ground. Fl. 9(1). Lee: Hoffmann 1920 (NEBC). Adventive from farther west.

KOCHIA

*__K. scoparia__ (L.) Roth ex Schrad. Summer-cypress
Uncommon. Waste ground. Fl. 9(1-2). Hoffmann 1916 (NEBC). Introduced from Eurasia.

SALSOLA

*__S. collina__ Pallas
Uncommon, in waste areas. Fr. 8(1). Pittsfield: Ahles (78382) 1973 (MASS). Introduced from Europe.

*__S. kali__ L. Russian Thistle
(**S. kali** L. var. **caroliniana** (Walt.) Nutt.)
Uncommon. Waste areas. Fr. 8. Pittsfield: Ahles (78469) 1973 (MASS). Introduced from Eurasia.

AMARANTHACEAE

AMARANTHUS

*__A. albus__ L. Tumbleweed
Uncommon. Waste areas. Fr. 8(1). Great Barrington, Pittsfield.

*__A. blitoides__ S. Wats. Tumbleweed
Uncommon. Waste areas. Fr. 8(1). Pittsfield: Ahles (78394)

1973 (MASS). Adventive from farther west.

***A. hybridus** L. Green Amaranth
Occasional. Roadsides, abandoned fields. Fr. 8-9. New Marlboro. Introduced from Eurasia.

***A. lividus** L. Livid Amaranth
(**A. blitum** L.)
Uncommon. Wet soil at edge of backwater marsh. Fr. 9. Sheffield. Introduced from tropics. First discovered in the county by the author in 1989.

***A. retroflexus** L. Pigweed
Occasional. Roadsides. Fr. 9. Sheffield. Introduced from Eurasia.

PORTULACACEAE

CLAYTONIA

C. caroliniana Michx. Carolina Spring Beauty
Common. Cool, moist woods, often rich woods. Fl. 4(3)-5(2). Common in the northern half of the county, often whitening the forest floor when in flower. On roadsides at summit of Mt. Greylock.

C. virginica L. T Narrow-leaved Spring Beauty
Rare. On rich, moist soil of floodplain, edge of pasture. Fl. 5(2). Sheffield.

PORTULACA

***P. oleracea** L. Purslane, Pusley
Common. Cultivated and waste soil. Fl. 8-9. Introduced from Europe.

MOLLUGINACEAE

MOLLUGO

***M. verticillata** L. Carpetweed
Frequent. Cultivated and disturbed sandy soil. Fl. 8. Adventive from tropical America.

CARYOPHYLLACEAE

ARENARIA

***A. serpyllifolia** L. Thyme-leaf Sandwort
Frequent. Open dry calcareous soil or sandy soil. Fl. 5(4). Introduced from Europe.

CERASTIUM

C. arvense L. Field Chickweed
Occasional. Open rocky woods, serpentine ledge. Fl. 5(3). Florida, Great Barrington, Stockbridge, West Stockbridge. Mt. Washington.

***C. fontanum** Baumg. Common Mouse-ear Chickweed
(***C. vulgatum** L.)
Common. Gardens, lawns, roadsides and fields. Fl. 5(3). Introduced from Europe.

C. nutans Raf. Nodding Chickweed
Uncommon. Shaded ledges. Fl. 6(2). Great Barrinton, West Stockbridge: Hoffmann 1919 (NEBC).

DIANTHUS

***D. armeria** L. Deptford Pink
Common, in dry fields, roadsides. Fl. 6(3). Introduced from Europe.

***D. barbatus** L. Sweet William
Uncommon. Roadside escape. Fl. 7(1). Florida: Hoffmann 1909 (NEBC). Introduced from Europe.

***D. deltoides** L. Maiden Pink
Common. Dry fields, roadsides. Fl. 6(1)-8(1). Introduced from Europe.

LYCHNIS

***L. flos-cuculi** L. Ragged-robin
Common. Moist, open fields. Fl. 6(1-2). Introduced from Europe.

MINUARTIA

M. michauxii (Fenzl) Farw. SC Rock Sandwort
(**Arenaria stricta** Michx.)
Rare. Dry calcareous ledges. Fl. 6(1). Sheffield.

MOEHRINGIA

M. lateriflora (L.) Fenzl Grove Sandwort
Occasional. Sandy, often alluvial soil. Fl. 5(4)-6(3).

M. macrophylla (Hook.) Fenzl **E** Large-leaved Sandwort
Rare. On shaded serpentine outcrops. Fl. 5(3). Cheshire, Florida.

MYOSOTON

***M. aquaticum** (L.) Moench Giant Chickweed
(**Stellaria aquatica** (L.) Scop.)
Common. Moist cultivated soil, alluvial soil. Has spread extensively since 1922. Introduced from Europe.

PARONYCHIA
P. canadensis (L.) Wood Forked Chickweed

Uncommon. Dry, rocky south-facing slopes. Fl. 8(1). Mt. Washington.

*__P. fastigiata__ (Raf.) Fern. Hairy Forked Chickweed
Uncommon. Dry roadsides and gravel bars. Fl. 8-9. Egremont. Great Barrington. Probably adventive from farther south. First discovered in the county by the author in 1988.

PETRORHAGIA

*__P. prolifer__ (L.) Ball & Heyw. Childing Pink
(*__Dianthus prolifer__ L.)
Uncommon. Dry open sandy soil in old field with limestone outcrops. Fl. 8(4). Sheffield. Introduced from Europe.

SAGINA

*__S. japonica__ (Sw.) Ohwi Japanese Pearlwort
Occasional. Gravel margins of roads, driveways. Probably spreading rapidly. Fl. 6(3). Williamstown, Egremont, Florida. Introduced from Asia. First discovered in the county by the author in 1990.

*__S. procumbens__ L. Pearlwort
Common. Moist disturbed soil, lawns and pastures, waste areas. Fl. 6(3)-9(2). Introduced from Eurasia.

SAPONARIA

*__S. officinalis__ L. Soapwort, Bouncing-bet
Common. Moist soil along brooks, roadsides. Fl. 7(1)-9(2). Introduced from Europe.

SCLERANTHUS

*__S. annuus__ L. Knawel
Uncommon. Dry sandy soil, roadsides and pastures. Fl. 6(2)-10. Sheffield. Introduced from Europe.

SILENE

*__S. alba__ (P. Mill.) Krause White Campion
(**S. pratensis** (Raf.) Godr. & Gren.)
Common. Fields, roadsides. Fl. 5(4)-9. Introduced from Europe.

S. antirrhina L. Sleepy Catch-fly
Uncommon. Dry sandy soil or exposed limestone ledges. Fl. 7(1)-8(4). Sheffield.

S. caroliniana Walt. var. **pensylvanica** (Michx.) Fern. Wild Pink
Uncommon. Dry, south-facing slope above Bash-Bish Rd. in open woods, somewhat rich. Fl. 6(1). Mt. Washington. Plate 2F.

*__S. noctiflora__ L. Night-flowering Catch-fly
Occasional. A garden escape in woods and waste places. Fl. 6(2)-8(4). Cheshire, Lanesboro, North Adams, Pittsfield. Sheffield: Churchill 1920 (NEBC). Introduced from Europe.

*__S. vulgaris__ (Moench) Garke Bladder Campion
(**S. cucubalus** Wibel)
Common. Fields, roadsides, waste places, woods. Fl. 6(2)-8(4). Introduced from Europe.

SPERGULA

*__S. arvensis__ L. Corn-spurrey
Uncommon. Cultivated fields and waste ground. Fl. 8(1)-9(2). Sheffield: Churchill 1920 (NEBC). Introduced from Europe.

SPERGULARIA

*__S. rubra__ (L.) Presl & Presl Common Sand-spurrey
Uncommon. Waste areas, sandy lawns. Fl. 8(2). Pittsfield: Hoffmann 1914 (NEBC). Introduced from Europe.

*__S. marina__ (L.) Griseb. Saltmarsh Sand-spurrey
Uncommon. Sandy edge of highway, salted in winter. Williamstown. From coastal areas to the east. First discovered in the county by the author in 1993.

STELLARIA

S. borealis Bigelow ssp. **borealis** Northern Stitchwort
(**S. calycantha** (Ledeb.) Bong.)
Uncommon. Moss-covered rocks in springy seeps, small, cold streams and in cool, swampy woods. Fl. 6(1)-8(1). Florida, Hancock, Lanesboro. Mt. Washington.

*__S. graminea__ L. Lesser Stitchwort
Common. Fields, roadsides. Fl. 6(2)-9(3). Introduced from Europe.

S. longifolia Muhl. ex Willd. Long-leaved Stitchwort
Uncommon. On tussocks in open marshy pasture, swampy woods and damp thickets. Fl. 5(2)-6(4). Florida, Sheffield.

*__S. media__ (L.) Vill. Common Chickweed
Common. Disturbed and garden soil. Fl. 4(1)-9(1). Introduced from Europe.

POLYGONACEAE

FAGOPYRUM

*__F. esculentum__ Moench Buckwheat
Occasional. A field escape, on roadsides. Fl. 7(1)-9(1). Great

Barrington, Lanesboro. Introduced from Europe.

POLYGONELLA

***P. articulata** (L.) Meisn. Jointweed
Uncommon. Sandy, waste area. Fl. 9(1). Florida. Probably adventive from east. First discovered in the county by the author in 1988.

POLYGONUM

P. amphibium L. var. **emersum** Michx. Water Smartweed
Frequent. In shallow water in ponds.

P. amphibium L. var. **stipulaceum** (Coleman) Fern. Floating Smartweed
Frequent. Pondshores and swampy woods. Fl. 7(1).

***P. arenastrum** Bor. Doorweed
Uncommon. Sand bars in Housatonic River. Great Barrington: B. A. Sorrie & Weatherbee (NEBC).

P. arifolium L. var. **pubescens** (Keller) Fern. Halberd-leaved Tearthumb
Frequent. Marshes and wet woods. Fl. 7(4)-8(3).

***P. aviculare** L. Doorweed
Common. Lawns, fields. Fl. 7(3)-9(3). Introduced from Europe.

P. careyi Olney Smartweed
Uncommon. Muddy shores. Fl. 8(1). Becket: Hoffmann 1917 (NEBC).

***P. cespitosum** Blume var. **longisetum** (DeBruyn) Stewart Smartweed
Common. Along logging roads, roadsides. Fl. 7(1)-9(1). Introduced from Asia. First discovered in the county by B. A. Sorrie in 1988.

P. cilinode Michx. Fringed Bindweed
Common. Cool, mesic acidic woods, climbing over rocks and vegetation. Fl. 7(3).

***P. convolvulus** L. Black Bindweed
Common. Roadsides, railroad tracks, waste areas. Fr. 8(2). Introduced from Europe.

***P. cuspidatum** Sieb. & Zucc. Japanese Knotweed
Common. Abundant in floodplains, roadsides, waste areas, and an aggressive pest. Fl. 8(1). Introduced from Asia.

P. erectum L. Erect Knotweed
Uncommon. Roadsides and abandoned fields. Lee, Lenox, Williamstown. Lanesboro, Sheffield.

***P. hydropiper** L. Water-pepper
Frequent. Marshes and open wooded swamps. FL. 8(1). Introduced from Eurasia.

P. hydropiperoides Michx. Mild Water-pepper
Frequent. Shallow water of ponds and streams. Fl. 8(2).

P. lapathifolium L. var. **lapathifolium** Willow-weed
Frequent. River and stream banks. Fl. 8(4)-9(2). Possibly native.

P. lapathifolium L. var. **salicifolium** Sibthorp Willow-weed
Uncommon. Sandy shores. Fl. 8(4). Pittsfield: Hoffmann 1919 (NEBC). Possibly native.

***P. neglectum** Bess. Needle-leaf Knotweed
Uncommon. Moist, acidic ledges. Fl. 8(2). Mt. Washington: Hoffmann 1902 (NEBC). Introduced from Europe.

***P. nepalense** Meisn. Nepalese Knotweed
Uncommon. Gravel bars in Green River near NY-MA border. Fl. 8(2). Alford, Egremont. Introduced from Asia. First discovered in the county by the author in 1989.

***P. orientale** L. Prince's-feather
Uncommon, in waste ground. Fl. 9(2). Stockbridge: Hoffmann 1919 (NEBC). Introduced from Eurasia.

P. pensylvanicum L. var. **pensylvanicum** Smartweed
Common, in abandoned fields, waste ground. Fl. 8(2)-9(4).

***P. persicaria** L. Lady's Thumb
Common. Waste areas. Fl. 8(1). Introduced from Europe.

P. punctatum Ell. var. **confertiflorum** (Meisn.) Fassett Water Smartweed
Frequent. Marshes and along streams. Fl. 8(2).

***P. sachalinense** F. Schmidt Giant Knotweed
Uncommon. Roadsides. Fl. 8(1). Peru. Introduced from Asia. First discovered in the county by B. A. Sorrie in 1990.

P. sagittatum L. Tearthumb
Common. Climbing over vegetation in marshes and stream banks. Fl. 8(2).

P. scandens L. var. **scandens** Climbing False-buckwheat
Frequent. Moist soil, river banks. Fl. 9(1).

P. scandens L. var. **cristatum** (Engelm. & A. Gray) Gleason Winged Bindweed
Uncommon. Dry open rocky woods. Mt. Washington.

P. tenue Michx. Knotweed
Uncommon. Dry sandy open soil. Fl. 8(1). Alford. Sheffield: Walters 1915 (NHA).

P. virginianum L. Jumpseed
(**Tovara virginianum** (L.) Raf.)
Common. Moist soil in clearings, river banks.

RHEUM

***R. rhaponticum** L. Rhubarb
Occasional. Usually persisting at old house sites. Introduced from Europe.

RUMEX

***R. acetosa** L. Green Sorrel
Frequent. Fields. Fl. 6(1). Introduced from Europe.

***R. acetosella** L. Sheep Sorrel
Common. Old fields, cultivated soil. Fl. 5(4). Introduced from Europe.

***R. crispus** L. Curly Dock
Common. Waste areas, roadsides. Fl. 8(4). Introduced from Europe.

***R. longifolius** DC. Yard Dock
Uncommon. Wet meadows. Fl. 9(3). Otis: Hoffmann 1904 (NEBC). Introduced from Europe.

***R. obtusifolius** L. Bitter-Dock
Common. Marshes, moist roadsides. Fl. 9(2). Introduced from Europe.

R. orbiculatus Gray Great Water Dock
Frequent. Fens, alluvial soil, rich marshes in calcareous areas. Fl. 8(4).

***R. triangulivalvis** (Danser) Rechinger f. Willow-leaf Dock
Uncommon. Waste ground, abandoned beaver dam. Fl. 7-8(1). Pittsfield. Stockbridge, Williamstown. Adventive from farther north.

ACTINIDIACEAE

ACTINIDIA

***A. arguta** Sieb. & Zucc. Hardy Kiwi
Uncommon. Rich, moist soil. A high-climbing vine, climbing over trees and other vegetation, colonizing clearings, marble outcrops, riverbanks. Probably an escape from estate gardens as it was used as an arbor vine (Rehder, 1986). Fl. 6(2), Fr. 9. May also spread vegetatively by fragmentation. Several stations seen in Lenox, in W. Mtn. Rd. and E. Dugway Rd. area and in large wooded area remote from houses, all vigorous. Lenox, Stockbridge. Introduced from Asia. First discovered in the county in 1991 by the author.

ELATINACEAE

ELATINE

E. minima (Nutt.) Fisch. & Mey. Waterwort
Occasional. Sandy-bottomed shallow lakes and ponds.

***E. triandra** Schk. European Waterwort
Uncommon. Muddy-bottomed shallow ponds. Fr. 9(2). Great Barrington, New Marlboro. First discovered in the county by C. B. Hellquist in 1976.

CLUSIACEAE

HYPERICUM

H. ascyron L. **H** Great St. John's-wort
(**H. pyramidatum** Ait.)
Rare and historical. In river marshes, in alluvial soil. Fl. 7(4)-8(2). Stockbridge, Sheffield: Churchill 1920 (NEBC).

H. boreale (Britt.) Bickn. Northern St. John's-wort
Occasional. Peaty moist shores of ponds. Fl. 7(3)-8(2). Savoy.

H. canadense L. Canadian St. John's-wort
Frequent. Wet soil along streams, margins of ponds. Fl. 7(2-4).

H. ellipticum Hook. Pale St. John's-wort
Common. Open wet meadows, marshes and pond shores. Fl. 7(1-4).

H. gentianoides (L.) BSP. Pineweed
Uncommon. Open dry sandy soil. Fl. 7(1). New Marlboro. Sheffield.

H. majus (Gray) Britt. Greater St. John's-wort
Occasional. Peaty or muddy shores, mudflats of abandoned beaver ponds. Fl. 7(4)-8(3).

H. mutilum L. Dwarf St. John's-wort
Frequent. Marshes and wet meadows, along small streams. Fl. 8(1-2).

***H. perforatum** L. Common St. John's-wort
Common. Waste areas, roadsides, fields, railroad tracks. Fl. 6(4)-7(3). Introduced from Europe.

***H. prolificum** L. Shrubby St. John's-wort
Uncommon. Escaped from gardens, in moist, disturbed soil at edge of old estate. Fl. 6(4). Lenox. Native farther west. First discovered in the county by the author in 1987.

H. punctatum Lam. Spotted St. John's-wort
Common. Marshes and moist thickets. Fl. 7(3-4).

TRIADENUM

T. virginicum (L.) Raf. Marsh St. John's-wort
Common. Edges of bogs, old beaver ponds, marshes. Fl. 7(4)-8(1-2).

TILIACEAE

TILIA

T. americana L. Basswood
Common. Alluvial soil along rivers and in rich woods. Fl. 7(2).

MALVACEAE

ABUTILON

***A. theophrasti** Medic. Velvet-leaf
Frequent. Weed in abandoned fields. Fl. 9(1). Introduced from Asia.

ALCEA

***A. rosea** L. Hollyhock
(**Althaea rosea** (L.) Cav.)
Uncommon. Garden escape and occasionally self-sown. Fl. 8(1). Sheffield: Hoffmann 1919 (NEBC). Introduced from Eurasia.

HIBISCUS

***H. trionum** L. Flower-of-an-hour
Occasional. Cultivated soil. Fl. 7(4). Stockbridge, Sheffield, Williamstown. Introduced from Europe.

MALVA

***M. alcea** L. Hollyhock-mallow
Uncommon. Roadsides and in fields. Fl. 7(4). Great Barrington. Lee: Hoffmann 1920 (NEBC). Introduced from Europe.

***M. moschata** L. Musk-mallow
Occasional. Abandoned fields, roadsides. Fl. 7(3)-9(1). Lanesboro, Stockbridge, Williamstown. Introduced from Europe.

M. neglecta Wallr. Cheeses
Common. Gardens, waste ground. Fl. 6(1). Introduced from Europe.

SARRACENIACEAE

SARRACENIA

S. purpurea L. Pitcher-plant
Frequent. Sphagnum mat of bogs, more often on the Berkshire Plateau. Fl. 6(3-4).

DROSERACEAE

DROSERA

D. intermedia Hayne Spatulate-leaved Sundew
Occasional. In sphagnum mat in bare muddy edges of pools and boggy shores of ponds. Fl. 7(4). Mt. Washington, Otis, Savoy, Sheffield.

D. rotundifolia L. Round-leaved Sundew
Frequent, in sphagnum mat of bogs, damp sand and wet slopes on roadsides. Fl. 7(2)-8(1).

CISTACEAE

HELIANTHEMUM

H. bicknellii Fern. Frostweed
Uncommon. Dry sandy soil, sand plains. Fl. 7(1). Sheffield. Great Barrington: Churchill 1920 (NEBC).

H. canadense (L.) Michx. Frostweed
Uncommon. Dry, sandy soil along road in sand plain. Fl. 6(1). Great Barrington, Sheffield. Plate 3A.

LECHEA

L. intermedia Leggett ex Britt. Pinweed
Occasional. Dry, sandy soil, in clearings and fields. Florida, Great Barrington, Sheffield. Williamstown .

VIOLACEAE

There is much hybridization between many of the species of stemless blue violets. Taxonomy is also in flux.

VIOLA

V. affinis LeConte LeConte's Violet
Frequent. Grassy meadows, alder thickets. Fl. 5(3).

***V. arvensis** Murr. Wild Pansy
Occasional. Garden escape, in fields and roadsides. Fl. 4(3)-7(2). Sheffield. Stockbridge: Hoffmann 1919 (NEBC). Introduced from Europe.

V. blanda Willd. Sweet White Violet
Frequent. Moist rich woods. Fl. 5(1-2).

V. canadensis L. Canada Violet
Frequent. Cool moist rich woods. Abundant on slopes of Mt. Greylock. Fl. 5(1) and sporadically throughout summer.

V. conspersa Reichenb. Dog Violet
Frequent. Moist open wet meadows and roadsides. Fl. 4(4)-5(4).

V. cucculata Ait. Marsh Violet
(**V. obliqua** Hill)
Common. Open wet meadows, wet areas in woods. Fl. 5(2-4).

V. fimbriatula Sm. Fringed Violet
(**V. sagittata** Ait.)
Common. Dry old fields, banks and open woods, sandy soil. Fl. 4(4)-5(3).

V. incognita Brainerd Large-leaf White Violet
(**V. blanda** var. **palustriformis** Gray)
Frequent. Dry open woods. Fl. 4(4)-5(4).

V. lanceolata L. Lance-leaf Violet
Occasional. Open rocky shores, in crevices of rock border of boggy pond. Fl. 5(2). Lee. Mt. Washington.

V. macloskeyi Lloyd ssp. **pallens** (Banks ex DC.) M.S. Baker
(**V. pallens** (Banks ex DC.) Brainerd)
Swamp White Violet
Frequent. Cool springy open meadows, dripping wet mossy rocks along brooks. Fl. 4(4)-5(4).

V. nephrophylla Greene **E** Northern Bog Violet
Rare. On cool wet ledges, edges of higher altitude marshes. Above 610 m, mostly on Mt. Greylock and Berkshire Plateau. Fl. 5(4).

***V. odorata** L. English Violet
Uncommon. Moist rich soil. Fl. 4(2). Stockbridge, Williamstown. Introduced from Europe.

V. primulifolia L. Primrose Violet
Uncommon. Boggy edge of pond. Fl. 7(1). Otis, Savoy.

V. pubescens Ait. var. **eriocarpa** (Schwein.) Russell
(**V. eriocarpa** Schwein.)
Smooth Yellow Violet
Common. Moist open woods and edges. Fl. 4(4)-5(3).

V. pubescens Ait. var. **leiocarpa** (Fern. & Wieg.) Seym.
(**V. pensylvanica** Michx. var. **leiocarpa** (Fern. & Wieg.) Fern.)
Smooth Yellow Violet
Common. Moist, rich woods. Fl. 4(4)-5(4).

V. pubescens Ait. var. **peckii** House Downy Yellow Violet
Uncommon. Fl. 5(3-4). Adams, Mt. Washington: Knowlton & Schweinfurth 1919 (NEBC).

V. pubescens Ait. var. **pubescens** Downy Yellow Violet
Frequent. Rich moist soil. Fl. 4(4)-5(3).

V. renifolia A. Gray var. **renifolia** Kidney-leaf Violet
Occasional. Rich moist woods, near swamps. Fl. 4(4)-5(4). Egremont, Stockbridge, Great Barrington, Williamstown.

V. renifolia Gray var. **brainerdii** (Greene) Fern.
Frequent. Moist rich woods, near fens. Fl. 5(3)-6(2). Pittsfield: Sorrie 2189 1983 (NEBC).

V. rostrata Pursh Long-spurred Violet
Frequent. Moist rich woods. Fl. 4(4)-5(4). Plate 3B.

V. rotundifolia Michx. Early Yellow Violet
Common. Dry open woods, particularly favoring trail edges. Fl. 4(1)-5(2).

V. sagittata Ait. Arrow-leaved Violet
Uncommon. Acidic dry fields, banks. Fl. 4(4)-5(4). Cheshire, Lanesboro.

V. selkirkii Pursh ex Goldie Selkirk's Violet
Occasional. Cool rich woods. Fl. 4(4)-5(4). Generally in northern half of county, in cool ravines in southern.

V. septentrionalis Greene Northern Blue Violet
Frequent. Dry sandy soil, open woods, rich woods. Fl. 4(4)-5(1).

V. sororia Willd. Woolly Blue Violet
(**V. papilionacea** Pursh pro parte)
Common. Moist fields and woods. Fl. 5(1-4).

V. triloba Schwein. var. **triloba** Three-lobed Violet
(**V. palmata** L. var. **triloba** (Schwein.) Gingins)
Occasional. Dry calcareous rocky open woods. Fl. 5(1-4). Mt. Washington, Great Barrington, Sheffield.

V. triloba Schwein. var. **dilatata** (Ell.) Brainerd
(**V. palmata** var. **dilatata** Ell.) Three-lobed Violet
Occasional. Dry calcareous rocky open woods. Fl. 5(1-2). Great Barrington, Mt. Washington, Sheffield, Egremont.

CUCURBITACEAE

ECHINOCYSTIS

E. lobata (Michx.) Torr. & Gray Wild Cucumber
Common. River banks, hedgerows, disturbed areas. Can be aggressive, swarming over other vegetation. Fl. 8-9.

SICYOS

S. angulatus L. Bur Cucumber
Common. Climbing on vegetation in fields, hedgerows, floodplains. Fl. 8-9.

SALICACEAE

POPULUS

***P. alba** L. White Poplar
Occasional. Disturbed habitats. Escaped and spreading along roadsides, fields. Introduced from Europe.

P. balsamifera L. Balsam Poplar
Frequent. Rich, calcareous seepage areas, river banks. Hoffmann (1922) thought only trees at Hoosac Tunnel, Florida were indigenous, and that other populations had spread from planted trees. The species is listed by Dewey (in Field, 1829) for Williamstown. The species may have spread more widely into newly cleared areas and from planting since Hoffmann's time.

P. deltoides Bartr. ex Marsh. Cottonwood
Common. Floodplains, river banks and shores, disturbed areas along roads. Fl. 5(2).

P. grandidentata Michx. Big-tooth Aspen
Common. Disturbed woodlands, persisting in mature forests.

***P. nigra** L. Lombardy Poplar
Uncommon. An escape from cultivation. Pittfield: Hoffmann 1916 (NEBC). Introduced from Europe.

P. tremuloides Michx. Trembling or Quaking Aspen
Common. In many kinds of disturbed soil, along roads, old fields, waste areas, clearings. Forms large clones as it spreads from root sprouts. Fl. 3(4).

SALIX

***S. alba** L White Willow
Common. Along streams, gravel bars in rivers, alluvial soil. Fl. 5(1). Introduced from Europe.

***S. alba** X ***S. fragilis**
Occasional. Along streams. Williamstown

S. bebbiana Sarg. Beaked Willow
Common. Wet meadows, edges of woods, moist roadsides. Fr. 5(4)-6(2).

S. bebbiana X **S. sericea**
Uncommon. Florida: Hoffmann 1920 (NEBC).

S. candida Flügge ex Willd. Hoary Willow
Frequent. Shrub fens, calcareous wet meadows, borders of calcareous ponds. Fl. 5(1).

S. Xclarkei Bebb
(**S. candida** X **S. petiolaris**)
Uncommon. Sheffield: Hoffmann 1920 (NEBC).

S. discolor Muhl. Pussy-willow
Common. In shrub swamps, marshy borders of ponds. Fl. 3(4).

S. eriocephala Michx. Heart-leaved Willow
Common. Marshes, shrub swamps, edges of ponds and streams, ditches. Fl 5(1).

S. eriocephala X **S. sericea**
Uncommon, borders of streams, shores. Stockbridge: Hoffmann 1920 (NEBC).

***S. fragilis** L. Crack-willow
Frequent. Stream banks, in ditches along roads. Fl. 5(2). Introduced from Europe.

S. humilis Marsh. var. **humilis** Upland Willow
Occasional. Dry open woods and sand plains. Florida, New Marlboro, Sheffield. Great Barrington: Hoffmann 1920 (NEBC).

S. lucida Muhl. Shining Willow
Frequent. Marshes, shrub swamps. Fl. 5(2-4).

S. lucida X **S. serissima**
Uncommon. Calcareous marshes and fens. Fr. 7. Lenox, Stockbridge.

S. nigra Marsh. Black Willow
Common. Along streams and rivers. Fl. 5(1).

S. pedicellaris Pursh Bog Willow
Occasional, in bogs and open fens. Fl. 5(2). Monterey, Richmond, Pittsfield. Lanesboro.

***S. pentandra** L. Bay-leaf Willow
Uncommon. Ditches, damp fields, stream banks. Fl. 5(1). Great Barrington, Williamstown. Introduced from Europe.

S. petiolaris J. E. Smith Slender Willow
(Including **XS. subsericea** (Anderss.) Schneid., thought to be a variant of **S. petiolaris** (Argus, 1986).)
Occasional. Open wet meadows. Hinsdale, Lenox.

***S. purpurea** L. Basket Willow
Occasional. Along streams and moist roadsides. Lenox, Richmond, Savoy, Sheffield. Introduced from Europe.

S. sericea Marsh. Silky Willow

Common. Shrub swamps, borders of streams and ponds. Fl. 4(3-4).

S. serissima (Bailey) Fern. Autumn Willow
Frequent. In shrub fens and calcareous wet meadows. Fl. 5(4)-6(2). Fr. 7(2). In central valley region.

BRASSICACEAE

ALLIARIA

***A. petiolata** (Bieb.) Cav. & Grande Garlic Mustard
Common. Roadsides, floodplains, choking out native vegetation. Fl. 4(4). Not listed by Hoffmann (1922) but noted as having been observed in the county. Introduced from Europe.

ARABIDOPSIS

***A. thaliana** (L.) Heynh. Mouse-ear Cress
Uncommon. Disturbed open ground. Fl. 5(1). Williamstown. Noted by Dewey (1829) but not by Hoffmann (1922). Introduced from Europe.

ARABIS

A. canadensis L. Sickle-pod
Occasional. Dry calcareous open woods. Fl. 6(2). Great Barrington, New Marlboro, North Adams, Sheffield. Williamstown.

A. drummondii Gray Rock-cress
Uncommon. Silty floodplain and bank of Deerfield River in open thicket. Fl. 5(2). Florida.

A. glabra (L.) Bernh. Tower-mustard
Common. Open dry fields and open rocky soil. Fl. 6(2).

A. hirsuta (L.) Scop. var. **pycnocarpa** (Hopkins) Rollins Hairy Rock-cress
Occasional. Open or partially shaded calcareous ledges. Fl. 5(2). New Ashford, Sheffield.

A. laevigata (Muhl. ex Willd.) Poir. **T** Smooth Rock-cress
Rare. On shaded, calcareous ledges and outcrops. Fl. 5(1-3). Adams, Lenox, Mt. Washington, North Adams, Sheffield, Stockbridge, West Stockbridge.

A. lyrata L. **T** Lyre-leaf Rock-cress
Rare. On open or lightly shaded ledges, usually with some calcareous influence. Fl. 4(3)-5(4). Mt. Washington, Sandisfield, Sheffield, West Stockbridge. Plate 3C.

ARMORACIA

***A. rusticana** (Lam.) Gaertn., Mey. & Scherb. Horseradish
Uncommon. Railroad yard, waste areas. North Adams: Knowlton 1915 (NEBC). Introduced from Europe.

BARBAREA

***B. vulgaris** R. Br. Winter-cress
Common. Roadsides, fields, wet woods edges. Fl. 5(2)-6(1). Introduced from Europe.

BERTEROA

***B. incana** (L.) DC. Hoary Alyssum
Common. Dry fields, roadsides. Fl. 6(1)-7(1). Introduced from Europe.

BRASSICA

***B. juncea** (L.) Czern. Brown Mustard
Common. Fields and roadsides. Fl. 7(1). Introduced from Europe.

***B. nigra** (L.) Koch Black Mustard
Common. Alluvial soil, roadsides. Fl. 6(4)-7(3). Introduced from Europe.

***B. rapa** L. Field Mustard
(***B. campestris** L.)
Common. Moist soil in fields, roadsides. Fl. 6(1)-7(1). Introduced from Europe.

CAPSELLA

***C. bursa-pastoris** (L.) Medic. Shepherd's Purse
Common. Waste areas, roadsides. Fl. 5(2)-6(3). Introduced from Europe.

CARDAMINE

C. bulbosa (Schreb. ex Muhl.) BSP. Spring Cress
Uncommon. Wet soil along small streams in calcareous swamps. Fl. 5(3-4). Egremont, Great Barrington, Sheffield.

C. douglassii (Torr.) Britt. **E** Purple Cress
Rare. In rich alluvial soil, partly shaded. Fl. 5(1-2). Sheffield.

***C. impatiens** L. Bushy Rock-Cress
Uncommon. On open rocky ledges, riverbanks. Fl. & Fr. 6(2). Sandisfield. Introduced from Europe. First discovered in the state by the author in 1993.

C. parviflora var. **arenicola** (Britt.) O. E. Schultz. Small-flowered Bittercress
Uncommon. Crevices on open ledges, rocky slopes with very open woods. Fl. 5(3-4). Sandisfield, Mt. Washington, West Stockbridge.

C. pensylvanica Muhl. ex Willd. Pennsylvania Bittercress
Frequent. Wet soil in swamps, along streams, moist soil in clearings. Fl. 5(2-4).

***C. pratensis** L. var. **pratensis** Cuckoo-flower
Common. Moist lawns and pastures, wet soil along brooks, often growing in dense colonies covering large areas. Fl. 5(1)-6(1). Flowers pale purple or pink. Introduced from Europe.

C. pratensis L. var. **palustris** Wimm. and Grab. **T**
Fen Cuckoo-flower
Rare. Growing singly in pools of water in open or lightly shaded fens. Fl. 5(3). Flowers white.

CONRINGIA

***C. orientalis** (L.) Dumort. Treacle-mustard
Uncommon. Waste areas. Fl. 7. Lee: Leahey 1927 (MASS).

DENTARIA

D. Xanomala Eames
(**D. diphylla** X **D. laciniata**)
Uncommon. Rich woods. Fl. 4(4)-5(2). Sheffield: Walters 1915 (NEBC).

D. diphylla Michx. Toothwort
(**Cardamine diphylla** (Michx.) Wood)
Common. Rich woods, moist soil. Fl. 4(4)-5(2).

D. laciniata Muhl. ex Willd. Cut-leaved Toothwort
(**Cardamine concatenata** (Michx.) Sw.)
Frequent. Moist rich woods. Fl. 4(3).

ERUCA

***E. vesicaria** (L.) Cav. Garden-rocket
Uncommon. Dry sterile fields. FL. 6(3). Pittsfield. Introduced from Europe.

ERUCASTRUM

***E. gallicum** (Willd.) O. E. Schultz French Rocket
Frequent. Waste places. Fl. 8-9. Introduced from Europe.

ERYSIMUM

***E. cheiranthoides** L. Wormseed-mustard
Common. Waste areas, fields. Fl. 5(4)-8. Introduced from Europe.

HESPERIS

***H. matrionalis** L. Dame's Rocket
Common. Moist roadsides and floodplains. Fl. 5(4)-7(4). Introduced from Europe.

IBERIS

***I. amara** L. Candytuft
Uncommon. Waste areas, sandy soil along railroad. Fl. 7(1). Great Barrington: Schweinfurth 1915 (NEBC). Introduced from Europe.

LEPIDIUM

***L. campestre** (L.) R. Br. Field Cress
Occasional, on roadsides and in fields. Fl. 5(4). Introduced from Europe.

***L. densiflorum** Schrad. Wild Peppergrass
Uncommon. Dry roadsides. Fl. 6(1). Native status in question, but weedy tendency.

L. virginicum L. Wild Peppergrass
Common. Waste areas, garden weed. Fl. 6(2).

NASTURTIUM

***N. microphyllum** (Boenn.) Reichenb. Watercress
(**Rorippa microphylla** (Boenn ex Reichenb.) Hyl.)
Occasional. Along cold streams. Fl. 6. Lanesboro, Stockbridge. Introduced from Europe.

***N. officinale** R. Br. Watercress
(**Rorippa nasturtium-aquaticum** (L.) Hayek)
Common. Springs and streams. Fl. 6. Introduced from Europe.

RAPHANUS

***R. raphanistrum** L. Wild Radish
Occasional. Roadsides, waste areas. Fl. 6(1)-7(4). Williamstown. Introduced from Europe.

***R. sativus** L. Garden Radish
Uncommon. Waste areas, a garden escape. Fl. 6-9(1). Lenox: Hoffmann 1912 (NEBC). Introduced from Europe.

RORIPPA

R. palustris (L.) Bess. var. **fernaldiana** (Butters & Abbe) Stuckey
(**R. islandica** (Oeder) Borbas ssp. **fernaldiana** (Butters & Abbe) Hultén)
Yellow Cress
Frequent. On silty banks of Housatonic River, as a weed in moist fields and disturbed areas near the river. Fl. 7(1)-8.

R. palustris (L.) Bess. ssp. **hispida** (Desv.) Jonsell
Yellow Cress

Uncommon. Waste ground. North Adams, Stockbridge, Lanesboro: Churchill 1916 (NEBC).

***R. sylvestris** (L.) Bess. Creeping Yellow Cress
Occasional. River banks and meadows. Fl. 9-10. Sheffield. Introduced from Eurasia.

SINAPSIS

***S. arvensis** L. Charlock
(**Brassica kaber** (DC.) D. C. Wheeler)
Common. Fields. Fl. 7-8. Pittsfield: Hoffmann 1902 (NEBC). Introduced from Europe.

SISYMBRIUM

***S. altissimum** L. Tumble-mustard
Frequent. Roadsides and waste areas. Fl. 6-9. Great Barrington: Hoffmann 1920 (NEBC). Introduced from Europe.

***S. officinale** (L.) Scop. Hedge-mustard
Common. Roadsides and waste areas. Fl. 6-8. Introduced from Europe.

THLASPI

***T. arvense** L. Field Pennycress
Occasional. Disturbed soil. Fl. 5(2). Great Barrington, Williamstown. Introduced from Europe.

CLETHRACEAE

CLETHRA

C. alnifolia L. Sweet Pepper-bush
Uncommon. Moist shore of Lake Marguerite. Fl. 8. Sandisfield.

ERICACEAE

ANDROMEDA

A. glaucophylla Link Bog Rosemary
(**A. polifolia** L. var. **glaucophylla** (Link) DC.
Occasional. Shrub border of bogs and boggy portions of fens. Fl. 5(4).

ARCTOSTAPHYLOS

A. uva-ursi (L.) Spreng. Bearberry
Uncommon. Open exposed rocky ridges, mainly on southern Taconics. Fl. 5(4). East Mt., Great Barrington; West Stockbridge Mt.; Race Mt., Mt. Washington.

CHAMAEDAPHNE

C. calyculata (L.) Moench var. **angustifolia** (Ait.) Rehder Leatherleaf
Frequent. Shrub border of bogs and edges of ponds. Fl. 5(1).

EPIGAEA

E. repens L. Trailing Arbutus, Mayflower
Frequent. Dry somewhat open acidic woods, at edges of clearings, trails. Fl. 4(2).

GAULTHERIA

G. hispidula (L.) Muhl. ex Bigelow Creeping Snowberry
Occasional. Creeping over moss in spruce-fir bog forest, often on moss-covered logs. Mt. Greylock Range and Berkshire Plateau. Fl. 5(4).

G. procumbens L. Wintergreen, Checkerberry, Teaberry
Common. Acidic open woods. Fl. 8(1).

GAYLUSSACIA

G. baccata (Wang.) Koch Black Huckleberry
Common. Dry open acidic woods and open ridges. Fl. 5(4).

KALMIA

K. angustifolia L. Sheep Laurel, Lambkill
Common. Clearings in dry acidic woods, or at edges of bogs. Fl. 6(1)-7(1).

K. latifolia L. Mountain Laurel
Common. Dry open rocky acidic woods and ridges, often forming impenetrable thickets. Most common on the Berkshire Plateau and southern third of county. Fl. 6(1-4).

K. polifolia Wang. Bog Laurel
Occasional. Shrub border and open sphagnum mat of bogs. Fl. 5(4). Otis, Savoy, Washington.

LEDUM

L. groenlandicum Oeder Labrador Tea
Occasional to frequent. Edges of bogs, boggy swamps and rarely, open mountain ridges. Fl. 6(1).

LYONIA

L. ligustrina (L.) DC. Male-berry
Common. Swamps, wet meadows, open fields. Fl. 6(4).

RHODODENDRON

R. canadense (L.) Torr. Rhodora

Uncommon. Bogs and swamps. Fl. 5(2). Not noted by Hoffmann. On May 27, 1938, E. H. Lincoln, in a letter to M. L. Fernald, stated that "the men working in the mountains near here brought me a bunch of them, found in the town of Washington." This letter is filed at NEBC. No specimens can be found documenting any sites. Habitat still exists, and this species may eventually be relocated. Listed by Dewey (in Field, 1829) for Pittsfield.

R. maximum L. **T** Great Laurel
Rare. In wooded swamps, forming colonies. Fl. 6(4)-7(1). Pittsfield, Richmond. Never common, it was almost extirpated by gardeners digging plants for large estates (Hoffmann, 1922). Plate 3D.

R. prinophyllum (Small) Millais Mountain Azalea
(**R. roseum** (Loisel.) Rehd.)
Common. Open acidic oak-blueberry woods, rocky ridges, edges of swamps. Fl. 5(4)-6(3).

VACCINIUM

V. angustifolium Ait. Lowbush Blueberry
Common. Open acidic woods, open rocky ledges, old fields on mountain ridges. Fl. 5(2)-6(1). Berries most sought after for picking.

V. corymbosum L. Highbush Blueberry
(Including **V. atrococcum** (Gray) Heller)
Common. Open wooded swamps, edges of ponds and high open ridges. Fl. 5(3-4).

V. macrocarpon Ait. Large Cranberry
Occasional. Creeping over sphagnum mat of bogs, boggy wet meadows. Fl. 7(1).

V. myrtilloides Michx. Velvet-leaf Blueberry
Occasional. Higher elevation cool acidic woods. Fl. 6(1).

V. oxycoccos L. Wren's Egg Cranberry
Uncommon. Creeping over open sphagnum mat of bogs. Fl. 6(2). Becket, Otis, Pittsfield, Savoy, Sheffield.

V. pallidum Ait. Early Sweet Blueberry
(**V. vacillans** Kalm ex Torr.)
Frequent. Open acidic oak woods on dry, south-facing slopes. Fl. 5(2).

V. stamineum L. Deerberry
Uncommon. Open woods on rocky south-facing slopes. Sheffield , Mt. Washington.

V. vitis-idaea L. ssp. **minus** (Loddiges) Hultén **E**
Mountain Cranberry
Rare. On exposed ledge, elevation 945 m. Fl. 6(2). Adams. First discovered in the state by C. Quinlan in 1982. Plate 3E.

PYROLACEAE

CHIMAPHILA

C. maculata (L.) Pursh Spotted Wintergreen
Occasional. Open acidic woods. Fl. 8(3). Florida, Hinsdale, Lanesboro, New Marlboro, Sheffield.

C. umbellata (L.) Bart. ssp. **cisatlantica** (Blake) Hultén
Pipsissewa
Frequent. Dry open oak woods. Fl. 7(1-3).

MONESES

M. uniflora (L.) Gray One-flowered Wintergreen
Uncommon. Moist woods often under white pine. Fl. 7(1-2) Egremont, North Adams, Sandisfield.

ORTHILIA

O. secunda (L.) House One-sided Wintergreen
(**Pyrola secunda** L.)
Occasional. Open dry rich woods. Fl. 7(3).

PYROLA

P. americana Sweet Round-leaf Pyrola
(**P. rotundifolia** L. var. **americana** (Sweet) Fern.)
Occasional. Dry open rich woods. Fl. 7(2).

P. asarifolia Michx. var. **purpurea** (Bunge) Fern. **E**
Pink Pyrola
Rare. In forested fens. Fl. 7(1-4). Stockbridge. Pittsfield: Sorrie & PBW 2185 1983 (NEBC).

P. chlorantha Sweet Green Shinleaf
(**P. virens** Schweig. ex Schweig. & Koerte)
Occasional. Moist, somewhat rich woods. Fl. 6(4).

P. elliptica Nutt. Shinleaf
Common. Open woods. Fl. 7(2-3).

MONOTROPACEAE

MONOTROPA

M. hypopithys L. Pinesap
Occasional. Woods, often under evergreens. Fl. 8(1).

M. uniflora L. Indian Pipes
Common. Moist woods. Fl. 7(1).

PRIMULACEAE

LYSIMACHIA

L. ciliata L. Fringed Loosestrife
(**Steironema ciliatum** (L.) Baudo)
Common. Borders of moist woods, stream banks, moist thickets. Fl. 7(2).

***L. nummularia** L. Moneywort
Common. In moist soil, a lawn and garden weed, in wet woods. Fl. 6(2). Introduced from Europe.

L. Xproducta (Gray) Fern.
(**L. terrestris** X **L. quadrifolia**)
Uncommon. Border of damp woods, near pond. Fl. 7(3). Great Barrington, Sandisfield, Washington.

L. quadrifolia L. Whorled Loosestrife
Common. Dry, open acidic woods. Fl. 6(3).

L. terrestris (L.) BSP. Swamp-candles
Frequent. Circumneutral marshes and swamps, wet meadows. Fl. 7(1-3).

L. thyrsiflora L. Tufted Loosestrife
(**Naumbergia thyrsiflora** (L.) Duby)
Occasional. Calcareous marshes. Fl. 6(2).

***L. vulgaris** L. Garden Loosestrife
Occasional. River marshes along Housatonic River, wet meadows, marshy border of ponds. Fl. 7(3)-9. Introduced from Europe.

PRIMULA

***P. anisodora** Balf.f. & Forrest Anise Primrose
Uncommon. Seepage area near old pump site. Fl. 5(3). Alford: Sorrie 2374 1984 (NEBC). First discovered in the state by B. A. Sorrie.

TRIENTALIS

T. borealis Raf. Starflower
Common. Moist or dry, usually acidic woods, clearings. Fl. 5(2)-6(1).

HYDRANGEACEAE

HYDRANGEA

***H. arborescens** L. Wild Hydrangea
Uncommon. Moist woods along trail in Canoe Meadows Sanctuary, a former estate. Fl. 8(2). Pittsfield. Adventive from farther west or escaped from cultivation.

PHILADELPHUS

***P. pubescens** Loisel. Mock-orange
Occasional. Along roadsides. Fl. 6(1). A garden escape, from farther south.

GROSSULARIACEAE

RIBES

R. americanum Mill. Wild Black Currant
Frequent. Alluvial soil on river banks, calcareous swamps. Fl. 5(3).

R. cynosbati L. Dogberry, Prickly Gooseberry
Common. Rich woods and hedgerows. Fl. 4(4)-5(2).

R. glandulosum Grauer Skunk Currant
Frequent. Cool moist acidic woods, more often in higher altitudes on Berkshire Plateau, Mt. Greylock Range. Fl. 5(1-3).

R. hirtellum Michx. Smooth Gooseberry
Frequent. Fens and calcareous swamps. Fl. 5(3).

R. lacustre (Pers.) Poir. **SC** Bristly Black Currant
Rare. On wet rocks along brooks in ravines at high elevations where there is some calcareous influence, and occasionally in moist rich soil. Most abundant on Mt. Greylock Range. Fl. 5(4). Adams, Florida, Hancock, New Ashford, North Adams, Savoy, Williamstown, Windsor.

***R. odoratum** Wendl.f. Golden Currant
Uncommon. A garden escape. Fl. 6(1). New Marlboro: Hoffmann 1912 (NEBC). Introduced from Europe.

***R. rubrum** L. Garden Red Currant
Common. Moist roadsides and river banks. Fl. 4(4). Introduced from Europe.

R. triste Pallas **SC** Swamp Red Currant
Rare. In cool fens, calcareous wet soil in rich woods, springy areas and cold brooks. Fl. 4(4)-5(3). Cheshire, Hinsdale, Lanesboro, New Ashford, North Adams, Pittsfield, Richmond, Stockbridge, West Stockbridge, Williamstown.

CRASSULACEAE

PENTHORUM

P. sedoides L. Ditch Stonecrop
Common. On muddy margins of marshes and ponds. Fl. 7(3).

SEDUM

***S. acre** L. Stonecrop
Uncommon, on calcareous ledges. Fl. 6(2)-7. Sheffield. Introduced from Eurasia, a garden escape.

***S. spurium** Bieb. Sedum
Uncommon. Garden escape on limestone rocks in dry field. Sheffield. Introduced from Eurasia.

***S. telephium** L. Orpine, Live-forever
(Including **S. purpureum** (L.) Schultes)
Common. On roadsides, hedgerows. Rarely flowers. Introduced from Eurasia.

S. ternatum Michx. Sedum, Mountain Stonecrop
Uncommon. On moist rocks, base of steep, rocky hill. Fl. 5(2). Sheffield.

SAXIFRAGACEAE

CHRYSOSPLENIUM

C. americana Schwein. ex Hook. Golden Saxifrage
Frequent. Open springy areas, wet places in woods, borders of small streams. Fl. 4(4)-5(2).

MITELLA

M. diphylla L. Mitrewort
Frequent. On calcareous rocks in rich woods, moist rich woods. Fl. 4(4)-6(2).

M. nuda L. Mitrewort
Occasional. On mossy hummocks in forested fens, swampy woods. Fl. 5(2-4).

PARNASSIA

P. glauca Raf. Grass-of-Parnassus
Frequent. Open, calcareous seepage areas and wet meadows. Fl. 8(2)-9(4).

SAXIFRAGA

S. pensylvanica L. Swamp Saxifrage
Frequent. In circumneutral forested swamps. Fl. 5(4).

S. virginiensis Michx. Early Saxifrage
Frequent. Calcareous open or lightly shaded ledges. Fl. 4(4)-5(4). Plate 3C.

TIARELLA

T. cordifolia L. Foamflower
Common. Moist rich woods, moist soil along borders of streams. Fl. 5(2)-6(2).

ROSACEAE

AGRIMONIA

A. gryposepala Wallr. Agrimony
Common. Borders of woods, hedgerows, clearings. Fl. 7(3).

A. parviflora Ait. **E** Small-flowered Agrimony
Rare. In alluvial soil, wet disturbed pastures along stream into Housatonic R. and moist meadow. Fl. 8(3-4). Fr. 8-10(2). Sheffield. First discovered in the state by the author in 1989. Since it would be difficult for earlier botanists to miss this tall agrimony, it probably migrated north from Connecticut after Hoffmann's time.

A. pubescens Wallr. **T** Downy Agrimony
Rare. In open dry calcareous woods, clearings and along trails in calcareous areas. Fl. 7(3-4). Egremont, Lenox, Sheffield, West Stockbridge.

A. striata Michx. Agrimony
Common. Open woods, borders of woods and thickets. Fl. 7(3)-8(2).

AMELANCHIER

A. arborea (Michx.f.) Fern. var. **arborea**
Shadbush, Serviceberry
Common. Dry woods borders, pastures, hedgerows, open woods Fl. 4(4)-5(2).

A. arborea (Michx.f) Fern. var. **laevis** (Wieg.) Ahles
Smooth Shadbush
Common. Woods borders, pastures, old fields, hedgerows, wooded swamps. Possibly blooming slightly later, and reaching higher elevations and moister habitats, generally, than the preceding. Fl. 4(4)-5(3).

A. bartramiana (Tausch) Roem. **T** Bartram's Shadbush
Rare. In montane boreal forest, summit of Mt. Greylock, and boggy boreal forest above 914 m (3000′). Also sporadically on exposed ledges on Hoosac Range, and in spruce-fir swamps on Berkshire Plateau at 609 m (2000′), and in one lowland fen. Fl. 5(2)-6(1). Adams, New Ashford, North Adams, Peru. Plate 3F.

A. canadensis (L.) Medicus Thicket Shadbush
Uncommon. Hancock: R. C. Bean 1919 (NEBC).

A. Xneglecta Egglest.
(**A. bartramiana** X **A. arborea** var. **laevis**)
Occasional. Calcareous marshes and rich moist woods. Fl. 5(2). Lanesboro.

A. sanguinea (Pursh) DC. **SC** Round-leaf Shadbush

Rare. On open rocky ridge and mountain summits, with some calcareous influence. Fl. 5(2). Alford-West Stockbridge, Great Barrington, Sheffield, Stockbridge.

A. stolonifera Wieg. Thicket-shadbush
Occasional. Open acidic rock ledges and ridgetops and rarely in sandy floodplain of Deerfield R. Forming colonies. Fl. 5(2-3). Florida, Great Barrington, North Adams, West Stockbridge, Williamstown.

A. stolonifera Wieg. forma **micropetala** (Robins.) Rehd.
Uncommon. Open quartzite ridge and mountain summit. Fl. 5(2). Smaller (not over 0.5 m) than preceding, with few branches and solitary terminal inflorescences. Petals 3-4 mm long. Williamstown.

ARONIA

A. arbutifolia (L.) Pers. Red Chokeberry
Frequent. Bogs and fens, also rocky ridges. Fl 5(4).

A. melanocarpa (Michx.) Ell. Black Chokeberry
Common. Open rocky woods and ridges, old fields and swamps. Fl. 5(2)-6(1).

A. prunifolia (Marsh.) Rehd. Purple Chokeberry
Occasional. Swamps, bogs and in open moist clearings at higher elevations. Fl. 6(2). This taxon, intermediate between the two preceding taxa are thought to be the result of ancient and ongoing hybridization between the the two preceding taxa (Hardin, 1973).

CRATAEGUS

The classification and nomenclature for **Crataegus** follows that of Gleason (1952). Identifications were made using both Gleason and Seymour (1982). No attempt is made here to cite all the synonyms of each species or group. **Crataegus** spp. occupy a variety of habitats, usually old fields and pastures or other anthropogenic habitats, and each species does not seem to be restricted to any one habitat, therefore habitats are not described.

C. brainerdii Sarg. Hawthorn, Thorn Apple
Common. Fl. 5(4). Mt. Washington.

C. chrysocarpa Ashe Round-leaf Hawthorn
Fl. 5(4). New Marlboro.

C. compta Sarg.
Mt. Washington: Knowlton & Schweinfurth 1919 (NEBC).

C. dodgei Ashe
Fl. 5(4). Egremont: Churchill & Schneider 1919 (NEBC).

C. flabellata (Bosc) Koch
Becket.

C. holmesiana Ashe
Adams: Knowlton & Bean 1915 (NEBC).

C. ideae Sarg.
Williamstown: Churchill 1900 (NEBC).

C. intricata Lange
Sheffield: Bean & Fernald 1919 (NEBC).

C. macrosperma Ashe
(**C. fretalis** Sarg.)
Mt. Washington: Knowlton & Schweinfurth 1919 (NEBC).

***C. monogyna** Jacq.
West Stockbridge. Introduced from Europe.

C. pedicellata Sarg.
West Stockbridge: Evans, Fernald & Knowlton 1919 (NEBC).

C. pruinosa (Wendl.f.) Koch
North Adams: Fernald & Long 1913 (NEBC).

C. punctata Jacq.
Common. Near rivers. Sheffield.

C. succulenta Schrad. ex Link. var. **macracantha** (Lodd.) Egglest.
(**C. macracantha** Lodd.)
Williamstown: Churchill 1900 (NEBC).

C. succulenta Schrad. ex Link. var. **succulenta**
Egremont: Hoffmann 1914 (NEBC).

DALIBARDA

D. repens L. Dewdrop
Occasional. Cool wooded swamps, shaded stream borders. Fl. 6(2-3). Clarksburg, Lenox, Hinsdale, Pittsfield, Savoy, Williamstown. Plate 4A.

FILIPENDULA

***F. rubra** (Hill) Robins. Queen-of-the-prairie
Occasional. Moist soil on roadsides and in marshes. Fl. 7(4). Adams, Stockbridge. Native farther west, escaped here.

***F. ulmaria** (L.) Maxim. Queen-of-the-meadow
Occasional. Marshes and moist soil. Fl. 7(2). Stockbridge. Most often seen in the double cultivated form. Introduced from Europe.

FRAGARIA

***F. ananassa** Dcne. Cultivated Strawberry

(**F. chiloensis** (L.) Dcne. **F. virginiana**)
Uncommon. Meadow borders and roadsides. Fl.6(1). Egremont: Hoffmann 1920 (NEBC).

F. vesca L. ssp. **americana** (Porter) Staudt Woodland Strawberry
Frequent, in moist, somewhat rich woods and moist, sandy soil under **Pinus strobus**. Fl. 5(4)-6(1). West Stockbridge.

*__F. vesca__ L. ssp. **vesca** European Strawberry
Frequent. Open habitat. Hancock: Hoffmann 1916 (NEBC).

F. virginiana Dcne. var. **virginiana** Wild Strawberry
Common. Meadows, pastures, roadsides. Fl. 4(4)-6(1).

GEUM

G. allepicum Jacq. Yellow Avens
(Including **G. allepicum** var. **strictum** (Ait.) Fern.)
Occasional. Woods' borders and in waste areas. Fl. 6(2)-7(3).

G. canadense Jacq. White Avens
Common. Moist woods and clearings. Fl. 6(2)-7(3).

G. laciniatum Murr. Floodplain Avens
Frequent. Moist alluvial soil. Riverbanks, muddy borders of brooks and woods. Fl. 6(3).

G. rivale L. Water Avens
Frequent. Calcareous wet meadows and wooded fens. Fl. 5(4)-6(4).

MALUS

*__M. pumila__ Miller Domestic Apple
(**M. domestica** Borkh.)
Common. Pastures, old fields, hedgerows, roadsides, young woods. Fl. 5(3-4). Introduced from Europe.

*__M. prunifolia__ (Willd.) Borkh. Chinese Crab-apple
(**Pyrus prunifolia** Willd.)
Uncommon. Roadsides. Fl. 5(3). Sheffield: Hoffmann 1920 (NEBC). Introduced from Asia.

*__M. sieboldii__ (Reg.) Rehd. Toringo Crab-apple
Uncommon. Roadside shrub borders. Fl. 6. Great Barrington. Becoming an aggressive weed in other parts of state. First discovered in the county by the author in 1990. Introduced from E. Asia.

PHYSOCARPUS

*__P. opulifolius__ (L.) Maxim. Ninebark
Frequent. Along river banks. Fl. 6(4). Not listed by Hoffmann. Probably adventive from farther west or escaped from cultivation.

POTENTILLA

*__P. argentea__ L. Silvery Cinquefoil
Common. Dry waste areas and fields. Fl. 6(1-3). Introduced from Eurasia.

P. arguta Pursh Tall Cinquefoil
Uncommon. Dry woods. Fl. 8(3). West Stockbridge: Hoffmann 1902 (NEBC).

P. canadensis L. Dwarf Cinquefoil, Five Fingers
Common. Dry fields and open woods. Fl. 5(3).

P. fruticosa L. Shrubby Cinquefoil
Frequent. Open calcareous wet meadows, shrub fens, dry limy rubble areas. Fl. 7-9. A useful indicator of calcareous conditions.

*__P. norvegica__ L. Rough Cinquefoil
Common. Moist or dry open disturbed areas. Fl. 6(4)-7(2). Possibly introduced from Europe.

P. palustris (L.) Scop. Marsh Cinquefoil
Occasional. Usually in calcareous marshes and fens but also in bogs. Fl. 6(3).

*__P. recta__ L. Sulfur Cinquefoil
Common. Dry disturbed areas. Fl. 6(2)-8. Introduced from Europe.

P. simplex Michx. Old-field Cinquefoil
Common. Dry open roadsides and fields. Fl. 5(2)-6.

P. tridentata Soland. ex Ait. Three-toothed Cinquefoil
Occasional. On open rocky ridges at higher elevations on Taconic and Hoosac Ranges, Ragged Mtn. in Greylock Range. Fl. 6(3)-7(4). Clarksburg, Florida, Mt. Washington, North Adams. Plate 4B.

PRUNUS

P. americana Marsh. Wild Plum
Uncommon. On roadsides. Fl. 5. New Marlboro: Hoffmann 1919 (NEBC).

*__P. avium__ (L.) L. Bird Cherry
Occasional. Hedgerows, pastures, overgrown woods. Fl 5(1). North Adams, Pittsfield, Sheffield. Introduced from Europe.

*__P. cerasus__ L. Sour Cherry
Uncommon. Roadsides, old house sites. Fl. 5(3). Stockbridge: Hoffmann 1914 (NEBC). Introduced from Europe.

***P. domestica** L. var**. insititia** (L.) Schneid. Garden Plum
(**P. insititia** L.)
Uncommon. Around old house sites. Fl. 5(2). New Marlboro: Hoffmann 1912 (NEBC). Introduced from Europe.

P. nigra Ait. Canada Plum
Occasional. Roadside thickets. Fl. 5(1). Sheffield,Williamstown.

P. pensylvanica L.f. Fire Cherry, Pin Cherry
Common. In burned areas, logging clearings, edges of woods, open woods. Fl. 5(2). Short-lived, a pioneer in disturbed areas, soon being over-topped by longer-lived trees.

P. pumila L. var. **susquehanae** (Willd.) Jaeger Sand Cherry
(**P. cuneata** Raf.; **P. susquehanae** Willd.)
Uncommon. Formerly on sandplain, now only on high open rocky ridges. Fl. 5(2). Mt. Washington

P. serotina Ehrh. Black Cherry
Common. Moist to dry woods, most common in disturbed woods, hedgerows. Fl. 6(1).

P. virginiana L. Choke-cherry
Common. Woods' borders, hedgerows, old fields, floodplains, summit of Mt. Greylock. Fl. 5(2)-6(1).

PYRUS

***P. communis** L. Pear
Occasional. Old pastures. Fl. 5(2). Introduced from Europe.

ROSA

R. acicularis Lindl. ssp. **sayi** (Schwein.) Lewis **E**
Northern Prickly Rose
Rare. On exposed calcareous ledge. Fl. 6(1). Lanesboro. First discovered in the state by the author in 1984. Plate 4C.

R. blanda Ait. Smooth Rose
Occasional. Dry open woods, gravelly river floodplain. Fl. 6(3-4). Florida.

***R. canina** L. Dog Rose
Uncommon. Rich pasture. Fl. 6(2-3). Williamstown . Introduced from Europe.

R. carolina L. Pasture or Carolina Rose
Frequent. Dry open wooded slopes and open ridges, often somewhat calcareous. Fl. 6(3)-7(1).

***R. cinnamomea** L. Cinnamon Rose
Uncommon. Roadsides and near old house sites. Fl. 7(1). Sheffield. Florida: Hoffmann 1909 (NEBC). Introduced from Eurasia.

***R. gallica** L. French Rose
Uncommon. Roadsides. Fl. 7(1-2). Great Barrington: Hoffmann 1920 (NEBC). Introduced from Europe.

***R. multiflora** Thunb. ex Murr. Multiflora Rose
Common. Pastures, roadsides, hedgerows, woods and calcareous ledges. Fl. 6(3). Introduced from Asia. Once recommended for hedges and wildlife, it is now one of the worst alien pests, choking out much native vegetation and creating impenetrable thorny tangles. Not noted by Hoffmann.

R. palustris Marsh. Swamp Rose
Common. Marshes, shrub swamps, edges of ponds. Fl. 7(2-3).

R. palustris Marsh. forma **inermis** (Regel) Lewis
Uncommon. Along streams and swamps. Fl. 7(2). Pittsfield, Stockbridge. Sheffield. This form, lacking any thorns on the floral branches, but with glandular hypanthium, may be hybrids between **R. palustris** and **R. blanda**. (See Lewis, 1957).

***R. rubiginosa** L. Sweetbriar, Eglantine
(**R. eglanteria** L.)
Uncommon. Moist rich pastures. Fl. 6(3). Williamstown. Introduced from Europe.

***R. rugosa** Thunb. Sea-beach Rose
Uncommon. Open disturbed areas. Fl.6(2). Pittsfield. Introduced from Asia.

***R. setigera** Michx. Prairie Rose
Occasional. Disturbed areas, roadsides, along RR tracks. North Adams, Stockbridge, Williamstown. From farther west.

RUBUS

Classification and identification of this genus follows Gleason's (1952) system of collective species.

R. allegheniensis Porter ex Bailey Blackberry
Common. Pastures, old fields, hedgerows, clearings. Fl. 6(2).

R. canadensis L. Smooth Blackberry
(Including var. **elegantulus** Blanchard)
Common. Clearings, along woods roads. More common in cool uplands. Fl. 6(2-4).

***R. cuneifolius** Pursh Sand Blackberry
Uncommon. Open pasture. Cheshire: Ahles 84366 1977 (MASS). Introduced from farther south and east.

R. enslenii Tratt. Southern Dewberry
Uncommon. Dry rocky woods. Fl. 5(4). Sheffield.

R. flagellaris Willd. Northern Dewberry

Common. Old fields, swamps and bogs. Fl. 6(1-4).

R. hispidus L. Swamp Dewberry, Running Blackberry
Common. Old fields, wet meadows, bogs. Fl. 7(1-3).

R. idaeus L. ssp. **strigosus** (Michx.) Focke Red Raspberry
(**R. strigosus** Michx.)
Common. Old fields, hedgerows, roadsides. Fl. 6(1-2).

***R. laciniatus** Willd. Cut-leaf Blackberry
Uncommon. Edge of pasture. Sheffield.. Introduced from Eurasia. First discovered in the county by the author in 1989.

R. occidentalis L. Black Raspberry
Common. Roadsides, clearings, old fields. Fl. 6(1-2).

R. odoratus L. Purple-flowering Raspberry
Common. Open woods, clearings, shaded ledges. Fl. 6(3)-8(4).

R. pensilvanicus Poir. ex Lam. High-bush Blackberry
Frequent. Wet or swampy ground, edges of brooks and swamps. Fl. 6(2)-7(1).

R. pubescens Raf. Dwarf Raspberry, Plumboy
Common. Moist or swampy woods, moist rich woods. Fl. 5 (3-4).

R. setosus Bigel. Bog Blackberry
Occasional. Wet meadows and moist pastures. Fl. 7(1-3). Lanesboro, Peru.

R. vermontanus Blanch. Vermont Blackberry
Uncommon. Moist soil. Thickets, roadsides. Fl. 7(2). Pittsfield, Sandisfield, Savoy. Becket Hoffmann 1920 (NEBC).

SANGUISORBA

S. canadensis L. Canadian Burnet
Uncommon. Along gravel edge of Deerfield River. Fl. 8(2-3). Florida. Plate 4D.

***S. minor** Scop. ssp. **muricata** Briq. Garden Burnet
Uncommon. Fen border. Fl. 6(3). Lee: Sorrie 3461 1986 (NEBC). Introduced from Eurasia.

SORBARIA

***S. sorbifolia** (L.) A. Br. False Spiraea
Occasional. Roadsides and in open woods. Fl. 7(1). Lenox, Sheffield. Introduced from Asia.

SORBUS

S. americana Marsh. Mountain-ash
(**Pyrus americana** (Marsh.) DC.)
Occasional. At elevations above 460 m (1500′) in open ridges, mountain summits and clearings in forest. Fl. 6(3)-7(1).

***S. aucuparia** L. European Mountain-ash
(**Pyrus aucuparia** (L.) Gaertn.)
Occasional. Disturbed areas, spreading from planted trees, usually seen as seedling or sapling. Lenox, Pittsfield. Introduced from Europe.

S. decora (Sarg.) Schneid. **E** Northern Mountain-ash
(**Pyrus decora** (Sarg.) Hyl.)
Rare. On open cold summits. Only on Mt. Greylock Range above 914 m (3000′). Fl. 6(1-2), a week to 10 days before **S. americana**. Adams. Plate 4E.

SPIRAEA

S. latifolia (Ait.) Borkh. Meadow-sweet
Common. Old fields, marshes, hedgerows, roadsides. Fl. 7(3)-8(4).

S. tomentosa L. Steeple-bush, Hardhack
Common. Old fields and pastures. Fl. 7(4)-8(4). White forms are occasionally seen.

WALDSTEINIA

W. fragarioides (Michx.) Tratt. **SC** Barren Strawberry
Rare. In open rich woods, alluvial soil and moist soil near swamps. Fl. 5(1-3). Great Barrington, Lanesboro, Pittsfield, Sheffield.

CAESALPINIACEAE

GLEDITSIA

***G. tricanthos** L. Honey-locust
Uncommon. Fl. 6(3). Egremont: Walters 1919 (NHA). Escaped from planted trees, native farther west and south.

SENNA

S. hebecarpa (Fern.) Irwin & Barneby **E** (**H**) Wild Senna
Rare and historical in Berkshire County. Moist soil. Hoffmann stated it is occasional in the southern part of the valley. Egremont, Sheffield,West Stockbridge: Walters 1916 (NHA).

FABACEAE

AMORPHA

***A. fruticosa** L. False Indigo
Uncommon. Stream borders, in swamps. Fl. 6(4). Cheshire, Stockbridge. Introduced from farther west.

AMPHICARPAEA

A. bracteata (L.) Fern. Hog Peanut
Common. Open moist, often rich woods. Fl. 7(4)-8(2).

APIOS

A. americana Medic. Groundnut
Frequent. Moist ground near marshes, river banks, clearings. Fl. 8(2-3).

BAPTISIA

B. tinctoria (L.) R. Br. var. **crebra** Fern. Wild Indigo
Occasional. Dry open woods, roadsides. Fl. 8(1-4). Great Barrington and southward.

CORONILLA

***C. varia** L. Crown Vetch
Occasional. Spread from plantings along roadsides. Fl. 6(4)-9. Introduced from Europe.

CROTALARIA

C. sagittalis L. Rattlebox
Uncommon. Dry sandy open ground. Fl. 8(1). Sheffield: Hoffmann 1899 (NEBC). Much of this habitat now overgrown.

DESMODIUM

D. canadense (L.) DC. Showy Tick-trefoil
Frequent. Dry open woods, river banks, railroad tracks. Fl. 7(4).

D. cuspidatum (Muhl. ex Willd.) Loud. Large-bracted Tick-trefoil
Occasional. Rocky open woods. Fl. 7(4)-8(4). North Adams, Sheffield. New Marlboro.

D. glutinosum (Muhl. ex Willd.) Wood Pointed-leaf Tick-trefoil
Frequent. Open rich woods and alluvial soil. Fl. 7(1)-8(2).

D. nudiflorum (L.) DC. Naked Tick-trefoil
Frequent. Dry rich open woods. Fl. 7(4)-8(3).

D. paniculatum (L.) DC. var. **paniculatum** Narrow-leaf Tick-trefoil
Occasional. Moist open rich woods and clearings. Fl. 8(2)-9(1). North Adams. More frequent from Lee south.

D. perplexum Schub. Dillen's Tick-trefoil
(**D. paniculatum** (L.) DC. var. **dillenii** (Darl.) Isely)
Occasional. Rich open woods, grows in same woods as preceding species (See Voss, 1985). Fl. 8(4). Sheffield

D. rotundifolium DC. Low Tick-trefoil
Occasional. Dry calcareous soil, open woods. Fl. 8(3). Egremont, New Marlboro. Sheffield.

LATHYRUS

***L. latifolius** L. Everlasting-pea
Occasional. Hedgerows, river banks. Fl. 7(4). Florida, Williamstown. Introduced from Europe. Not noted by Hoffmann.

***L. pratensis** L. Yellow Vetchling
Uncommon. Moist field. Fl. 6(3-4). Stockbridge. Introduced from Europe. Not noted by Hoffmann.

***L. sylvestris** L. Perennial Pea
Uncommon. Hedgerows, roadsides. 9(1). Florida. Introduced from Europe. Not noted by Hoffmann.

LESPEDEZA

L. capitata Michx. Roundhead Bush-clover
Occasional. Dry open sandy soil. Fl. 8(2-3). Florida, North Adams, Otis, Sheffield, Stockbridge, Wiiliamstown.

L. hirta (L.) Hornem. Hairy Bush-clover
Occasional. Dry open woods, dry clearings and edges. Fl. 8(2-3). Great Barrington, North Adams, Hancock, Williamstown.

L. intermedia (S. Wats.) Britt. Wand Bush-clover
Occasional. Dry rich soil, open woods and clearings. Fl. 8(2-3). Alford, Great Barrrington, Hancock, Sheffield, Stockbridge,Williamstown.

L. violacea (L.) Pers. Violet Bush-clover
Uncommon. Dry open calcareous woods and ledges. Fl. 8 (3). New Marlboro, Sheffield. Plate 4F.

LOTUS

***L. corniculatus** L. Bird's-foot Trefoil
Common. Roadsides, dry fields. Spread from plantings along roads. Fl. 7-9. Introduced from Eurasia.

LUPINUS

L. perennis L. Wild Lupine
Uncommon. Sandplain in pitch pine barrens. Fl. 5(2). Sheffield: Walters 1913 (NHA). Much of this habitat has been developed or become overgrown since fire has been reduced.

MEDICAGO

***M. lupulina** L. Black Medick
Common. Roadsides, waste areas. Fl. 7-9. Introduced from Eurasia.

***M. sativa** L. Alfalfa
Frequent. Fields, spread from agricultural plantings. Fl. 7-8. Introduced from Eurasia.

MELILOTUS

***M. alba** Desr. ex Lam. White Sweet-clover
Common. Abundant along roads and in waste areas. Fl. 6(4)-7(4). Introduced from Eurasia.

***M. officinalis** (L.) Pallas Yellow Sweet-clover
Frequent. Roadsides, waste areas. Common in southern part of county, spreading north. Fl. 6-8. Introduced from Europe.

ROBINIA

***R. pseudoacacia** L. Black Locust
Common. Disturbed habitats, hedgerows, fields, stream banks. Spread from plantings. Fl. 6(1-2). Introduced from farther south.

***R. viscosa** Vent. ex Vauq. Clammy Locust
Occasional. Escaping to roadsides from plantings. Fl. 6. Stockbridge, Pittsfield. Introduced from farther south.

TRIFOLIUM

***T. arvense** L. Rabbit's-foot Clover
Common. Dry roadsides. Fl. 7(3)-9. Introduced from Europe.

***T. aureum** Pollich Hop-clover
Common. Roadsides, waste areas. Fl. 6(4)-7(4). Introduced from Europe.

***T. campestre** Shreb. Low Hop-clover
Frequent. Pastures, on roadsides. Fl. 5(1)-6(4). Introduced from Europe.

***T. hybridum** L. Alsike Clover
Common, in fields and roadsides. Fl. 6-8. Introduced from Europe.

***T. incarnatum** L. Scarlet Clover
Uncommon. Orchard. Fl. 7(3). Sheffield: Walters 1915 (NHA). Probably non-persistent. Introduced from Europe.

***T. pratense** L. Red Clover
Common. Fields and roadsides. Fl. 7-8. Introduced from Europe.

***T. repens** L. White Clover
Common. Lawns, pastures, fields. Fl. 7-8. Introduced from Europe.

VICIA

***V. cracca** L. Bitter Vetch
Common. Fields, roadsides. Fl. 6-7. Probably introduced from Europe.

***V. sativa** L. ssp. **sativa** Common Vetch
Uncommon. Roadsides and river banks. Fl. 9. Sheffield. Florida: Hoffmann 1904 (NEBC). Introduced from Europe.

***V. tetrasperma** (L.) Moench Slender Vetch
Uncommon. Fields. Fl. 6(2)-7(2). Sheffield. Introduced from Eurasia. First discovered in the county by the author in 1989.

***V. villosa** Roth ssp. **villosa** Hairy Vetch
Uncommon, in fields and roadsides. Fl. 6(4). Sheffield: Hoffmann 1901 (NEBC). Introduced from Eurasia.

ELAEAGNACEAE

ELAEAGNUS

***E. angustifolia** L. Russian Olive
Common. Roadsides, hedgerows, fields. Fl. 5(4). Introduced from Eurasia. Not listed by Hoffmann.

***E. umbellata** Thunb. Autumn Olive
Occasional. Roadsides and in fields. Fl. 6. Introduced from E. Asia. Widely planted and becoming an aggressive weed in open areas. Not listed by Hoffmann.

HALORAGACEAE

MYRIOPHYLLUM

M. farwellii Morong **E** Farwell's Water Milfoil
Rare. Acidic boggy pond. Fr. 9. Savoy. First discovered in the state by the author in 1992.

***M. heterophyllum** Michx. Water Milfoil
Uncommon. Acidic ponds, Upper and Lower Goose Pond, 1000 Acre Lake. Lee-Tyringham, New Marlboro. Adventive from farther east and south, and may become weedy. First discovered in the county by the author in 1986.

M. humile (Raf.) Morong Water Milfoil
Uncommon. Acidic ponds on Berkshire Plateau. Savoy. First discovered in the county by the author in 1988.

M. sibiricum Komarov Water Milfoil
(**M. exalbescens** Fern.)
Frequent. Circumneutral or acidic ponds. Declining where **M. spicatum** occurs (Crow & Hellquist, 1983).

***M. spicatum** L. Eurasian Milfoil
Common. Abundant in alkaline waters where it impedes boating and swimming. Several methods are used to remove it including harvesting stems, herbiciding and water level drawdown. Introduced from Eurasia. Not listed by Hoffmann.

M. tenellum Bigel. Slender Water Milfoil
Uncommon. Acidic ponds, in sandy shallow water. Otis, Savoy. First discovered in the county by Jerry Jenkins in 1970.

M. verticillatum L. **T** Water Milfoil
Rare. Small peaty pools near central pond of lake basin graminoid fen. Stockbridge. First discovered in the county by the author in 1990.

PROSERPINACA

P. palustris L. Mermaid-weed
Frequent. Mud in shallow water of marshes, ponds and streams. Fl 7(2-3).

LYTHRACEAE

DECODON

D. verticillatus (L.) Ell. Water-willow
Occasional. Margins of bogs and rarely marshes. Fl. 7(4).

LYTHRUM

***L. alatum** Pursh Winged Loosestrife
Uncommon. Calcareous, sometimes disturbed, wet meadows. Fl 7(3-4). Lenox, Pittsfield, Stockbridge, Williamstown. Adventive from farther west.

***L. salicaria** L. Purple Loosestrife
Common. Abundant in wet meadows, marshes. A weedy pest crowding out other vegetation. Fl. 8-9. Introduced from Eurasia.

THYMELAEACEAE

DAPHNE

***D. mezereum** L. Daphne
Occasional. Usually in moist rich woods. Fl. 4(2). Introduced from Eurasia.

DIRCA

D. palustris L. Leatherwood, Wicopy
Occasional to frequent. Rich moist woods, where calcareous rock is near or at the surface. Fl. 4(3)-5(1).

ONAGRACEAE

CIRCAEA

C. alpina L. Dwarf Enchanter's Nightshade
Frequent. Cool moist woods, usually near wet areas. Fl. 7(1).

C. Xintermedia Ehrh.
(**C. alpina** X **C. lutetiana**)
Occasional. Clearings, in more open situations than preceding species. Fl. 7(1). Great Barrington.

C. lutetiana L. ssp. **canadensis** (L.) Ashers & Magnus
(**C. quadrisulcata** (Maxim.) Franch. & Sav.) var. **canadensis** (L.) Hara. Enchanter's Nightshade
Common. Woods, clearings, along trails, often weedy around houses. Fl. 7(1)-8(3).

EPILOBIUM

E. angustifolium L. Fireweed
Occasional. Open higher elevation ridges, power lines and mountain summits. Fl. 7-8. Mt. Greylock Range, a few Taconic Range summits, Dalton.

E. ciliatum Raf. ssp. **ciliatum** Willow-herb
(**E. adenocaulon** Hausskn.)
Frequent. River and stream banks, border of marshes. Fl. 9.

E. ciliatum Raf. ssp. **glandulosum** (Lehm.) Hoch & Raven
(**E. glandulosum** Lehm.) Northern Willow-herb
Frequent. Cool wet woods and along streams. Fl. 8(1-3).

E. coloratum Biehl. Purple-leaf Willow-herb
Common. Open moist disturbed soil, marshes. Fl. 8(3-4).

***E. hirsutum** L. Hairy Willow-herb
Occasional. Rich wet soil, roadsides, edges of marshes. Fl. 8(3). Introduced from Eurasia.

E. leptophyllum Raf. Willow-herb
Occasional. Wet meadows, marshes and bogs. Fl. 8.

E. palustre L. Marsh Willow-herb
Uncommon. Bogs and wet meadows. Fl. 7(3). Sheffield, Stockbridge.

E. strictum Muhl. ex Spreng. Downy Willow-herb
Occasional. Calcareous wet meadows. Fl. 7(4)-8(2).

GAURA

***G. biennis** L.
Gaura
Uncommon. In fields. Fl. 8. Lee: Hoffmann 1904 (NEBC). Adventive from farther south.

LUDWIGIA

L. palustris (L.) Ell. Water Purslane
Common. Mud in shallow water and margins of ponds and marshes. Fl. 8(3-4).

OENOTHERA

O. biennis L. Common Evening-Primrose
(Including **O. muricata** L.)
Common. Fields, roadsides, edges of woods. Fl. 6(2)-8(3).

O. perennis L. Sundrops
Frequent. Edges of woods, streams, moist fields. Fl.6(3)-7(4).

NYSSACEAE

NYSSA

N. sylvatica Marsh. Black Gum, Tupelo
Uncommon. Wet acidic sandy soil, wooded swamps. Fr. 8(1). Sheffield.

CORNACEAE

CORNUS

C. alternifolia L.f. Pagoda Dogwood
Common. Open woods, margins of woods, hedgerows. Fl. 6(1).

C. amomum Mill. ssp. **amomum** Silky Dogwood
Common. Margins of marshes and swamps. Fl. 7(1-3).

C. amomum Mill. ssp. **obliqua** (Raf.) J. S. Wilson Pale Dogwood
Occasional. Moist fields, edges. Fl. 7(1-3).

C. canadensis L. Bunchberry, Dwarf Cornel
Occasional. Under moist conifer woods on the Berkshire Plateau and Greylock Range, hummocks in swamps and fens in the valley. Fl. 5(4)-6(3).

C. florida L. Flowering Dogwood
Occasional. Open dry somewhat calcareous woods on south-facing slopes. Quite abundant on south-facing slope of Pine Cobble and East Mts. in North Adams and Williamstown in open oak woods and sporadic in southern third of county. Fl. 5(2-3). Plate 5A.

C. racemosa Lam. Gray Dogwood
(**C. foemina** Mill. ssp. **racemosa** (Lam.) J. Wilson)
Common. Old fields, drier edges of marshes, forming large patches. Fl. 6(3) and sporadically through summer.

C. rugosa Lam. Round-leaf Dogwood
Frequent. Edges of rich woods, on calcareous ledges. Fl. 6(3).

C. sericea L. ssp. **sericea** Red Osier Dogwood
(**C. stolonifera** Michx.)
Common. Wet fields, borders of marshes. Fl. 6(1-3).

SANTALACEAE

COMANDRA

C. umbellata (L.) Nutt. Bastard-toadflax
Occasional. Sandy soil, open woods, exposed rock outcrops. Fl. 5(4)-6(2). Mt. Washington, Sheffield, Williamstown. A hemiparasite on many different species (Voss, 1985).

VISCACEAE

ARCEUTHOBIUM

A. pusillum C. Peck **SC** Dwarf Mistletoe
Rare. A parasite found almost always on **Picea mariana**, causing a "witch's broom" growth on the host, and eventually killing it. Becket, Hinsdale, New Marlboro, Otis, Peru. Plate 5B.

CELASTRACEAE

CELASTRUS

*__C. orbiculata__ Thunb. Oriental Bittersweet
Common. In woods, hedgerows, roadsides. An aggressive destructive weed, climbing over vegetation, and winding around trees, strangling them. Fl. 6(3). Introduced from Asia. Not listed by Hoffmann.

C. scandens L. American Bittersweet
Occasional. Calcareous soil, fences and hedgerows, borders of woods and streams, exposed calcareous rocks as in marble quarries. Fl. 6(1-3). Similar to the preceding but not as high-climbing or aggressive.

EUONYMUS

*__E. alata__ (Thunb.) Sieb. Winged Euonymus
Frequent. Open woods. Fl. 6(1). Becoming more common and a potential aggressive weed. Not listed by Hoffmann . Introduced from Asia.

*__E. atropurpurea__ Jacq. Burning-bush, Wahoo
Occasional. Open woods, riverbanks. Fl. 6(3). Stockbridge. Sheffield. Native farther west.

*__E. europaea__ L. Spindle-tree
Uncommon. Hedgerows, borders of fields. Fl. 6(1-3). Stockbridge, Williamstown. A small tree, 5 m tall. Introduced from Eurasia.

AQUIFOLIACEAE

ILEX

I. laevigata (Pursh) Gray — Smooth Winterberry
Uncommon. In bog forest or forested fen under canopy of **Picea rubens** or **Larix laricina**. Fl. 6(4). Pittsfield, New Marlboro. Not noted by Hoffmann.

I. montana Torr. & Gray ex Gray **T** — Mountain Winterberry
Rare. Moist open rocky woods. Fl. 6(1). Mt. Washington. Northeasternmost location of this more southern species.

I. verticillata (L.) Gray — Winterberry
Common. Swamps and marshes. Fl. 6(3).

NEMOPANTHUS

N. mucronatus (L.) Loesener ex Koehne — Mountain Holly
Frequent. Edges of bogs and open rocky woods, open mountain ridges. Fl. 5(2-4).

BUXACEAE

PACHYSANDRA

*__P. terminalis__ Sieb. & Zucc. — Pachysandra
Uncommon. Open woods. Fl. 6. Sheffield: <u>Sorrie</u> <u>et</u> <u>al</u>. <u>2314</u> 1983 (NEBC). A garden escape. Introduced from Asia.

EUPHORBIACEAE

ACALYPHA

A. gracilens Gray — Three-seeded Mercury
Uncommon. Moist roadside. Fl. 8(3-4). Monterey.

A. rhomboidea Raf. — Three-seeded Mercury
Common. Waste areas, roadsides. Fl. 8(2-4).

CHAMAESYCE

C. maculata (L.) Sm. — Spotted Spurge
(**Euphorbia maculata** L.)
Common. Waste areas, roadsides, railroad tracks. Fl. 8(1-3).

C. nutans (Lag.) Sm. — Eyebane
(**Euphorbia preslii** Guss.)
Common. Dry waste areas, roadsides. Fl. 9(1-2).

C. vermiculata (Raf.) House — Hairy Spurge
(**Euphorbia vermiculata** Raf.)
Frequent. Dry soil on roadsides, waste areas. Fl. 8.

EUPHORBIA

***E. cyparissias** L. — Cypress Spurge
Common. Often calcareous soil, fields, roadsides. Fl. 5(1)-8. Introduced from Europe.

POINTSETTIA

*__P. dentata__ Michx.
Uncommon. Sheffield: <u>Sorrie</u> <u>4464</u> 1988 (NEBC). Native farther west and south.

RHAMNACEAE

CEANOTHUS

C. americanus L. — New Jersey Tea
Occasional. Open sandy or rocky south-facing banks in northern section, sandy roadsides and sandplains in southern. Fl. 6(4)-7(3). Plate 5C.

RHAMNUS

R. alnifolia L'Her. — Alder-leaf Buckthorn
Occasional. Calcareous swamps and fens. Fl. 5(2).

***R. cathartica** L. — Common Buckthorn
Common. Hedgerows, open woods, rock outcrops. Fl. 6(1). Introduced from Eurasia. An aggressive weed, it has spread widely since 1922.

***R. frangula** L. — Smooth Buckthorn
Common. Wet meadows, bogs and swamps, open woods. Fl. 6(2). Introduced from Eurasia. An aggressive weed, it has spread widely in a few years. Not listed by Hoffmann.

VITACEAE

PARTHENOCISSUS

P. quinquefolia (L.) Planch. ex DC. — Virginia Creeper
(Including **P. inserta** (Kerner) Fritsch)
Common. Hedgerows, open woods on rocks, borders of streams. Fl. 6(3)-7(1).

VITIS

V. aestivalis Michx. var. **argentifolia** (Munson) Fern. — Summer Grape
Occasional. Hedgerows, open rocky woods, south-facing slopes. Fl. 6(4)-7(1). More common in southern section of county.

V. labrusca L. — Fox Grape
Occasional. Rocky river banks and hedgerows. Fl. 6(4). Florida, New Marllboro, Otis, Sandisfield.

V. riparia Michx. — Riverbank Grape

Common. River banks, hedgerows, fences. Fl. 6(1-4).

LINACEAE

LINUM

***L. catharticum** L. Fairy-flax
Uncommon. Abundant along sandy gravel edge of unused gravel road, Moran Wildlife Management Area. Windsor. First discovered in the state by the author in 1993. From Europe.

L. medium (Planch.) Britt. var. **texanum** (Planch.) Fern. **T** Southern Yellow Flax
Rare. In open, dry limestone rubble. Fl. 8(3-4). Sheffield. First discovered in the county by the author in 1988.

L. sulcatum Ridd. **H** Grooved Flax
Rare and historical. In dry soil. Sheffield: Walters 1915 (NEBC).

***L. usitatissimum** L. Flax, Linseed
Uncommon, Open disturbed areas. Now a rare garden escape. Fl. 8. Stockbridge: Hoffmann 1902 (NEBC); noted in Williamstown, 1986 by PBW. Introduced from Eurasia.

L. virginianum L. Wild Yellow Flax
Occasional. Open dry calcareous meadows and rocky slopes. Fl. 7(4)-8(1). New Ashford, Sheffield, Williamstown.

POLYGALACEAE

POLYGALA

P. paucifolia Willd. Gay-wings, Fringed Polygala
Frequent. Open dry woods. Fl. 5(1)-6(1). Adams, Pittsfield and more commonly in south county.

P. polygama Walt. Bitter Milkwort
Uncommon. Dry open gravelly bank. Fl.7(2-4). West Stockbridge. First discovered in the county by B. A. Sorrie in 1990. Listed by Dewey (in Field, 1829), but not previously documented by a specimen.

P. sanguinea L. Purple Milkwort
Occasional. Dry sandy open fields and sandpits. Fl. 7(4)-9(4).

P. senega L. **H** Seneca Snakeroot
Rare and historical. River terraces, border of marsh. Fl. 5(4). Egremont: Churchill & Schneider 1919 (NEBC).

P. verticillata L. Whorled Milkwort
(Including varieties)
Occasional. Dry usually calcareous soil in open pastures, powerlines. Fl. 8(1)-9(4).

STAPHYLEACEAE

STAPHYLEA

S. trifolia L. Bladdernut
Occasional. Alluvial soil, dry calcareous soil, riverbanks, open woods.Fl. 5(2-4). Pittsfield south, but most common in Sheffield and Great Barrington.

HIPPOCASTANACEAE

AESCULUS

***A. hippocastanum** L. Horse-chestnut
Occasional. Moist soil. Fl. 5(3). Introduced from Europe.

ACERACEAE

ACER

***A. ginnala** Maxim. Amur Maple
Uncommon. Old fields. Fl. 5(4). Williamstown. Introduced from Asia. First discovered in the county by the author in 1989.

A. negundo L. Box-elder, Ash-leaf Maple
Common. Alluvial soil, floodplains, river banks, hedgerows in moist soil, moist disturbed soil. Fl. 5(1). Hoffmann considered it indigenous along the Housatonic River, but spread from plantings elsewhere.

A. nigrum Michx.f. **SC** Black Maple
Rare. Alluvial soil in floodplains, rich soil below calcareous ledges. Fl. 5(3). Adams, Egremont, Great Barrington, New Marlboro, North Adams, Pittsfield, Lenox, Williamstown.

A. pensylvanicum L. Striped Maple, Moosewood
Common. Moist northern hardwood forests. Fl. 5(2-4).

***A. platanoides** L. Norway Maple
Frequent. Spreading on river banks, moist roadsides. Fl. 5(1). Introduced from Europe. Not noted by Hoffmann.

A. rubrum L. Red Maple, Swamp Maple
Common. Dry woods, old fields, swamps, bogs. Fl. 4(2-3).

A. saccharinum L. Silver Maple
Frequent. Alluvial soil in floodplains. Fl. 4(2). Most common along the Housatonic River.

A. saccharum Marsh. Sugar Maple
Common. Moist rich woods, moist woods on mountain slopes. One of the most important trees in the northern hardwoods forest, most abundant in richer sites. Fl. 5(1).

A. spicatum Lam. Mountain Maple

Common. Moist soil in ravines, cool wet woods. Fl. 6(1).

ANACARDIACEAE

RHUS

***R. aromatica** Ait. Fragrant Sumac
Uncommon. Thicket on roadside bank along Rte. 8. Cheshire. Although reported in Berkshire County by Dewey (in Field, 1829), the above station is either planted or escaped.

R. copallinum L. Dwarf Sumac, Shining Sumac
Occasional. Dry, south-facing open slopes, or sandy fields. Fl. 8(1-4). Monterey, North Adams, Sheffield, West Stockbridge. Williamstown.

R. glabra L. Smooth Sumac
Frequent. Open, disturbed areas. Fl. 7(4).

R. typhina L. Staghorn Sumac
Common. Open disturbed sites, roadsides, railroad ROWs, old fields. Fl. 6(4).

TOXICODENDRON

T. radicans (L.) Kuntze Poison Ivy
(**Rhus radicans** L.)
Common. Moist or dry soil, roadsides. Climbing by thick woody stems with aerial roots that cling to tree trunks. Fl. 6(2). Common only in southernmost towns.

T. rydbergii (Small ex Rydb.) Greene Poison Ivy
(**Rhus radicans** L. var. **rydbergii** (Small ex Rydb.) Rehd.)
Common. Swamps, dry fields, roadsides. A low shrub. Fl. 6(2).

T. vernix (L.) Kuntze Poison Sumac
Occasional. Calcareous swamps and seeps. Fl. 6(4).

RUTACEAE

PTELEA

***P. trifoliata** L. Wafer-ash, Hop-tree
Uncommon. Alluvial soil along Housatonic River. Fl. 6(1-4). A few small trees spread from planting. Stockbridge. Native farther west and south.

ZANTHOXYLUM

Z. americanum Mill. Prickly-ash, Toothache-tree
Common. Alluvial soil and dry calcareous slopes. Fl. 5(4). Common in Egremont, Great Barrington, Sheffield, one station in Williamstown.

OXALIDACEAE

OXALIS

O. acetosella L. Wood-sorrel
(**O. montana** Raf.)
Frequent. Cool moist woods. Fl. 6(3-)-7(1). Less common in southern half of county in the valley region.

***O. corniculata** L. Creeping Sorrel
Uncommon. Around gardens and greenhouses. Introduced from Europe.

O. dillenii Jacq. ssp. **dillenii** Yellow Sorrel
Occasional. Dry soil. Fl. 8(2-3). Stockbridge. Lanesboro: Churchill 1915 (NEBC).

O. dillenii Jacq. ssp. **filipes** (Small) Eiten. Sorrel
Occasional. Dry calcareous soil. Fl. 6(1). Hancock, West Stockbridge. Sheffield: Churchill 1919 (NEBC).

O. fontana Bunge Yellow Sorrel
(**O. europaea** Jordan)
Common. Waste and disturbed sites. Fl. 6(1)-8(3).

GERANIACEAE

ERODIUM

***E. cicutarium** (L.) L'Her. ex Ait. Storksbill
Uncommon. In gardens as a weed. Fl. 6-8. Stockbridge: Hoffmann 1920 (NEBC). Introduced from Europe.

GERANIUM

G. bicknellii Britt. Geranium, Cranesbill
Uncommon. Rocky woods and ledges. Fl. 6(1). Sheffield: Bean & Fernald 1919 (NEBC).

***G. columbinum** L. Long-stalk Cranesbill
Uncommon. Dry sandy somewhat calcareous old field. Fl. 8. Sheffield. Introduced from Europe.

G. maculatum L. Wild Geranium
Common. Open woods and borders, roadsides. Fl. 5(2)-6(2).

***G. molle** L. Dovesfoot Cranesbill
Uncommon. Lawns. Fl. 7. Stockbridge: Hoffmann 1916 (NEBC). Introduced from Europe.

***G. pusillum** L. Slender Cranesbill
Uncommon. Farmyard. Fl. 6. Stockbridge: Hoffmann 1916 (NEBC). Introduced from Eurasia.

G. robertianum L. Herb Robert
Common. Moist calcareous ledges, talus and rich woods. Fl. 5(3) and throughout summer. Mostly in valley region.

***G. sibiricum** L. Cranesbill
Uncommon. Roadside and waste area. First discovered in the county by J. Strauch in 1994. From Eurasia.

BALSAMINACEAE

IMPATIENS

I. capensis Meerb. Spotted Jewelweed
Common. Wet places, along streams, banks, alluvial marshes, roadside ditches, borders of woods. Fl. 8-9.

***I. glandulifera** Royle Purple Balsam
Occasional. Alluvial marshes and rich soil along streams. Vigorous and spreading in same habitat as native **Impatiens**. Fl. 8(3)-9(4). Hancock, Lee, Richmond, Stockbridge. Introduced from Asia. First discovered in the county by B. A. Sorrie in 1988.

I. pallida Nutt. Pale Jewelweed
Common. Rich soil in moist rich woods, wet ravines slopes, roadsides. Becoming more common and almost weedy. Fl. 8(1)-9(2).

ARALIACEAE

ARALIA

A. hispida Vent. Bristly Sarsaparilla
Frequent. Open rocky ridge tops, ledges, logging clearings, open woods. Fl. 6(4)-7(3).

A. nudicaulis L. Wild Sarsaparilla
Common. Circumneutral to acidic woods. Fl. 5(3).

A. racemosa L. Spikenard
Frequent. Cool moist soil along streams, roadsides, borders of woods. Fl. 7(4)-8(2).

PANAX

P. quinquefolius L. **SC** Ginseng
Rare. Moist rich woods where calcareous rock forms outcrops or talus. Fl. 7(1). Most central valley region towns. Still occasionally collected for the root. Plate 5D.

P. trifolius L. Dwarf Ginseng
Frequent. Moist woods, usually at least somewhat rich. Fl. 4(4)-5(2).

APIACEAE

AEGOPODIUM

***A. podagraria** L. Goutweed
Common. Roadsides, alluvial soil, borders of woods. Fl. 6(1-4). An aggressive persistent weed. Introduced from Europe.

AETHUSA

***A. cynapium** L. Fool's Parsley
Uncommon. Gravelly riverbank. Fl. 7(4)-8(1). Egremont. Introduced from Europe. First discovered in the county by the author in 1990.

ANGELICA

A. atropurpurea L. Purple-stem Angelica
Frequent. Open alluvial soil, marshes, borders of streams. Fl. 6(1-4).

A. venenosa (Greenway) Fern. **H** Hairy Angelica
Rare and historical. Dry open woods on south-facing slope. Fl. 8(2). Great Barrington: Hoffmann 1911 (NEBC).

ANTHRISCUS

***A. sylvestris** (L.) Hoffm. Chervil
Uncommon. Roadsides, borders of fields, becoming an aggressive weed. Fl. 5(3). Lenox, Stockbridge. Introduced from Europe. First discovered in the county by the author in 1986.

CARUM

***C. carvi** L. Caraway
Occasional. Uually in higher elevations, on open ridge tops, pastures, roadsides. Fl. 6-7(1). Adams, Cheshire, Florida, Savoy.

CICUTA

C. bulbifera L. Water-hemlock
Common. Marshes. Fl. 7. Bulblets produced later in summer.

C. maculata L. Poison-Hemlock, Spotted Cowbane
Frequent. Wet soil, in marshes and swamps, along streams, roadsides. Fl. 6(3).

CONIOSELINUM

C. chinense (L.) BSP. **SC** Hemlock-parsley
Rare. In cold fens, shaded calcareous seeps, in cool ravines in soil with a calcareous influence. Fl. 8(3)-9(1). Great Barrington, Hinsdale, Lee, Lenox, New Ashford, New Marlboro, Richmond, Sheffield, Stockbridge, Williamstown.

CONIUM

***C. maculatum** L. Poison-Hemlock
Uncommon. Edges of thickets, fields, near abandoned farms. Fl. 7(3). Cheshire, New Marlboro. West Stockbridge: Evans, Fernald & Knowlton 1919 (NEBC). Introduced from Europe.

CRYPTOTAENIA

C. canadensis (L.) DC. Honewort
Common. Moist soil in open woods. Fl. 6(2-4).

DAUCUS

***D. carota** L. Queen Anne's Lace, Wild Carrot
Common. Fields, roadsides, waste areas. Fl. 6-9. Introduced from Europe.

FALCARIA

***F. vulgaris** Bernh. Sickleweed
Uncommon. Dry sandy calcareous soil, old field. Fl. 8(4). Sheffield. Introduced from Europe. First discovered in the state by B. A. Sorrie and the author in 1989.

HERACLEUM

H. lanatum Michx. Cow-parsnip
(**H. maximum** Bartr.)
Occasional. Moist roadsides, especially on the Berkshire Plateau. Fl. 6(1-3).

HYDROCOTYLE

H. americana L. Water Pennywort
Common. Moist or wet soil along streams, in wet meadows, borders of woods. Fl. 7-8.

LEVISTICUM

***L. officinale** Koch Lovage
Uncommon. Persistent near old house sites. Fl. 7(4). Monterey. Mt. Washington: Knowlton & Schweinfurth 1919 (NEBC). Introduced from Europe.

OSMORHIZA

O. claytonii (Michx.) Clarke Sweet Cicely
Common. Circumneutral or rich deciduous woods. Fl. 5(4).

O. longistylis (Torr.) DC. Anise-root
Occasional. Moist rich woods, often along roads. Fl. 6(1-2). In southern half of county.

PASTINACA

***P. sativa** L. Wild Parsnip
Common. Roadsides, waste places. Fl. 6(2)-7(3). Introduced from Europe.

PIMPINELLA

***P. saxifraga** L. Burnet-saxifrage
Uncommon. Spreading vigorously along moist rich roadside. Fl. 8(3). Sheffield, Williamstown. Introduced from Europe. First discovered in the county by the author in 1988.

SANICULA

S. canadensis L. **T** Canadian Sanicle
Rare. Well-drained soil in moist rich open woods near calcareous outcrops. Fl. 8(2-4). Sheffield. First discovered in the county by the author in 1993.

S. gregaria Bickn. **T** Long-styled Sanicle
(**S. odorata** (Raf.) K. M. Pryer & L. R. Philippe)
Rare. In rich, moist soil, often alluvial, near streams and swamps. Fl. 6(4)-7(1). Fr. 7(3)-9(1). Great Barrington, Lanesboro, Sheffield, Stockbridge, West Stockbridge. Williamstown.

S. marilandica L. Sanicle, Black Snakeroot
Common. Moist deciduous woods. Fl. 6(1). Fr. 7(3).

S. trifoliata Bickn. Trefoil Sanicle
Occasional. Moist rich woods. Fl. 6(2). Fr. 7(1)-8(1).

SIUM

S. suave Walt. Water-parsnip
Common. Swamps, marshes, muddy stream borders. Fl. 8.

TORILIS

***T. japonica** (Houtt.) DC. Hedge-parsley
Uncommon. Thickets along roadsides and railroads. Pittsfield, Stockbridge. Introduced from Eurasia. First discovered in the county by H. E. Ahles in 1972.

ZIZIA

Z. aurea (L.) Koch Golden Alexanders
Common. Meadows, borders of woods, roadsides. Fl. 4(4)-6(2).

GENTIANACEAE

BARTONIA

B. virginica (L.) BSP. Bartonia
Uncommon. Moist open or lightly shaded humus or peaty soil. Fl. 7(4). Great Barrington, North Adams. Williamstown.

PLATE 1

A. **Ophioglossum pusillum** Adder's Tongue Fern

B. **Cryptogramma stelleri** Slender Rock-brake

C. **Polystichum braunii** Braun's Holly Fern

D. **Woodsia glabella** Smooth Woodsia B.A. Sorrie

E. **Asplenium Xebenoides** Scott's Spleenwort

F. **Asplenium montanum** Mountain Spleenwort

PLATE 2

A. **Nuphar pumila** Tiny Cow Lily

B. **Ranunculus hispidus** var. **hispidus** Hispid Buttercup

C. **Ranunculus longirostris** White Water-crowfoot

D. **Adlumia fungosa** Climbing Fumitory

E. **Corydalis sempervirens** Pale Corydalis

F. **Silene caroliniana** var. **pensylvanica** Wild Pink

PLATE 3

A. **Helianthemum canadense** Frostweed

B. **Viola rostrata** Long-spurred Violet

C. **Arabis lyrata** *(center)* Lyre-leaf Rock-cress
Saxifraga virginiensis *(right & left)* Early Saxifrage

D. **Rhododendron maximum** Great Laurel

E. **Vaccinium vitis-idaea** Mountain Cranberry

F. **Amelanchier bartramiana** Bartram's Shadbush

PLATE 4

A. **Dalibarda repens** Dewdrop

B. **Potentilla tridentata** Three-toothed Cinquefoil

C. **Rosa acicularis** ssp. **sayi** Northern Prickly Rose

D. **Sanguisorba canadensis** Canadian Burnet

E. **Sorbus decora** Northern Mountain-ash

F. **Lespedeza violacea** Violet Bush-clover

PLATE 5

A. **Cornus florida** Flowering Dogwood

B. **Arceuthobium pusillum** Dwarf Mistletoe

C. **Ceanothus americanus** New Jersey Tea

D. **Panax quinquefolius** Ginseng

E. **Gentiana andrewsii** Andrews's Bottle Gentian

F. **Gentiana quinquefolia** Stiff Gentian

PLATE 6

A. **Asclepias quadrifolia** Four-leaf Milkweed

B. **Nymphoides cordata** Floating Heart

C. **Hydrophyllum canadense** Broad Waterleaf

D. **Verbena simplex**
Narrow-leaved Vervain

E. **Blephilia ciliata** Downy Wood-mint

F. **Blephilia hirsuta**
Hairy Wood-mint

PLATE 7

A. **Penstemon hirsutus** Hairy Beard-tongue

B. **Veronica catenata** Water Speedwell

C. **Conopholis americana** Squawroot

D. **Linnaea borealis** Twinflower

E. **Lonicera hirsuta** Hairy Honeysuckle

F. **Megalodonta beckii** Water-marigold

PLATE 8

A. **Arisaema dracontium** Green Dragon

B. **Heteranthera dubia**
Water Star-grass

C. **Chamaelirium luteum** Fairy Wand

D. **Calopogon tuberosus**
Grass Pink

E. **Cypripedium calceolus** var. **parviflorum**
Small Yellow Ladyslipper

B.A. Sorrie

F. **Isotria verticillata** Large Whorled Pogonia

CENTAURIUM

***C. pulchellum** (Sw.) Druce — Centaury
Frequent. Along moist roadsides in gravel, in waste areas. Fl. 7(2)-8(3). Introduced from Europe. First noted in 1989, now probably in every town.

GENTIANA

G. andrewsii Griseb. **T** — Andrews's Bottle Gentian
Rare. Calcareous soil in open woods, river banks and meadows. Fl. 9. One population with white flowers. Alford, Sheffield: Sorrie 1817 1982 (NEBC). Plate 5E.

G. clausa Raf. — Closed Gentian
Frequent. Moist shaded stream banks, borders of woods. Fl. 9.

G. linearis Froel. — Narrowleaf Closed Gentian
Uncommon. Moist border of cool woods. Fl. 8(3). Florida. First discovered in the county by the author in 1987.

G. quinquefolia L. — Stiff Gentian
Occasional. Dry or moist calcareous soil, meadows and open woods, in open disturbed soil as in old roads and trails. Cheshire, Great Barrington, Lenox, Mt. Washington, Sheffield. Fl. 8(2)-9(4). Plate 5F.

GENTIANOPSIS

G. crinita (Froel.) Ma — Fringed Gentian
(**Gentiana crinita** Froel.)
Frequent. Rich soil, open wet meadows and pastures, dry calcareous rubble as in old quarries Fl. 9-10(1).

APOCYNACEAE

APOCYNUM

A. androsaemifolium L. — Spreading Dogbane
Common. Meadows, on roadsides. Fl. 7(4).

A. cannabinum L. var. **cannabinum** — Indian Hemp
Frequent. Rocky river and pond shores and banks. Fl. 6(4)-7(4).

A. cannabinum L. var. **hypericifolium** A. Gray — Dogbane
(**A. sibiricum** Jacq.)
Occasional. Moist alluvial soil, moist meadows, near streams. Fl. 6(3)-7(2). Lee, Lenox, Monterey, Sheffield.

A. medium Greene — Dogbane
(**A. androsaemifolium** X **A. cannabinum**)
Occasional. Meadows and river banks. Fl. 7(1). Florida, Great Barrington, Sheffield.

VINCA

***V. minor** L. — Common Periwinkle
Occasional. Roadsides, woods near houses. Fl. 5(4). Introduced from Europe.

ASCLEPIADACEAE

ASCLEPIAS

A. amplexicaulis Sm. — Blunt-leaf Milkweed
Uncommon. Dry sandy soil. Fl.7(4). Sheffield: Walters 1915 (NHA). Loss of much of the open dry sandplain habitat may explain scarcity.

A. exaltata L. — Poke Milkweed
Frequent. Open rocky woods, border of woods. Fl. 6(3-4).

A. incarnata L. var. **incarnata** — Swamp Milkweed
Common. Wet meadows, border of marshes. Fl. 7(1)-8(2).

A. incarnata L. var. **pulchra** (Ehrh. ex Willd.) Pers. — Swamp Milkweed
Uncommon. Borders of ponds and marshes. Fl. 7(3). North Adams, Sandisfield, New Marlboro: Hoffmann 1912 (NEBC). Found on Berkshire Plateau.

A. purpurascens L. **H** — Purple Milkweed
Rare and historical. In open sandy field. Fl. 6(4)-7(4). Stockbridge: Hoffmann 1920 (NEBC). Specimen has only mature pods which appear to be that of this species.

A. quadrifolia Jacq. — Four-leaf Milkweed
Uncommon. Dry calcareous soil, on open south-facing, dry rich slopes, in open woods. Fl. 6(2)-7(1). New Marlboro, Sheffield, Stockbridge, West Stockbridge, Williamstown. Plate 6A.

A. syriaca L. — Common Milkweed
Common. Dry soil, meadows, roadsides, waste areas. Fl. 6(4)-8(1).

A. tuberosa L. — Butterfly-weed
Uncommon. Sandy dry open soil. Fl. 7(4). Great Barrington. Sheffield: Churchill 1920 (NEBC). Dewey reported this species from Berkshire County (in Field, 1829). Most open sandplain habitat has been lost to development or succession.

CYNANCHUM

***C. nigrum** (L.) Pers. — Black Swallow-wort
Frequent. Climbing on fences and vegetation on roadsides and along streams. Fl. 6(4)-7(4). Introduced from Europe. Not noted by Hoffmann.

SOLANACEAE

DATURA

***D. stramonium** L. Jimsonweed
Uncommon. Waste areas, dumps. Fl. 7(3)-8. Sheffield: Hoffmann 1912 (NEBC). Noted but not collected in Williamstown by PBW. Introduced from Asia.

LYCIUM

***L. barbarum** L. Matrimony-vine
Uncommon. Pasture. Fl. 7(3). New Marlboro: Hoffmann 1912 (NEBC). Introduced from Europe.

***L. chinense** Mill. Chinese Matrimony-vine
Uncommon. Waste area near old house site. Fl. 9(4). Cheshire. Introduced from Asia. First discovered by the author in 1989.

PHYSALIS

P. heterophylla Nees var. **heterophylla**
Clammy Ground-cherry
Common. Dry soil, waste areas, garden weed. Fl. 7(2)-8.

***P. subglabrata** Mackz. & Bush Ground-cherry
(**P. virginiana** Mill var. **subglabrata** (Mackz. & Bush) Waterfall)
Uncommon, Waste areas. Fl. 7-8. Stockbridge: Hoffmann 1902 (NEBC). Native farther south.

SOLANUM

S. americanum Mill American Black Nightshade
Frequent. Clearings in woods. Fl. 8(2)-9(3).

***S. carolinense** L. Horse-nettle
Frequent. Fields and disturbed sites. Fl. 8(3). Lenox, Sheffield. Native farther south and west. Not listed by Hoffmann.

***S. dulcamara** L. Bittersweet Nightshade
Common. Climbing on vegetation at edge of ponds, marshes, moist roadsides. Fl. 6(3)-9. Introduced from Europe.

***S. nigrum** L. Black Nightshade
Frequent. Alluvial soil and moist fields. Fl. 7-8. Introduced from Europe.

CONVOLVULACEAE

CALYSTEGIA

C. sepium (L.) R. Br. ssp. **americana** (Sims) Brummit
(**Convolvulus sepium L.** var. **americana** Sims)
Hedge-bindweed
Common. Climbing on vegetation in fields, fences, disturbed sites. Fl. 6(4)-8.

C. spithamaea (L.) Pursh ssp. **spithamaea H** Low Bindweed
(**Convolvulus spithamaeus** L.)
Rare. In open sandy soil. Fl. 6(4). Sheffield, New Marlboro: Hoffmann 1912 (NEBC). Habitat has become scarce due to development and succession.

CONVOLVULUS

***C. arvensis** L. Field Bindweed
Occasional. Waste areas, along railroads. Fl. 7(3). Introduced from Eurasia.

CUSCUTA

C. compacta Juss. ex Choisy Dodder
Occasional. Fields, clearings. Climbing on and parasitic on other vegetation. Fl. 8. West Stockbridge. First discovered in the county by the author in 1989.

***C. epithymum** (L.) L. Clover-dodder
Uncommon. Sandisfield: Walters 1914 (NEBC). Parasitic on other vegetation. Introduced from Europe.

C. gronovii Willd. ex Schultz Dodder
Common. Fields and roadsides, climbing on and parasitic on other vegetation. Fl. 8-9.

C. pentagona Engelm. Field-dodder
Uncommon. Climbing on and parasitic on other vegetation in dry open sandy fields. Fl. 8(3). Sheffield. First discovered in the county by B. A. Sorriein 1989.

MENYANTHACEAE

MENYANTHES

M. trifoliata L. var. **minor** Raf. Buckbean, Bogbean
Occasional. In water at edge of bog mat in shrub zone and in pools in fens. Fl. 5(4)-8.

NYMPHOIDES

N. cordata (Ell.) Fern. Floating Heart
Uncommon. Shallow water in acidic boggy ponds. Fl. 7(3). Becket, Otis, Mt. Washington. At present found in only one pond. Plate 6B.

POLEMONIACEAE

PHLOX

***P. paniculata** L. Perennial Phlox

Uncommon. Roadsides. Fl. 7-8. New Marlboro, Pittsfield. Native to farther west and south.

***P. subulata** L. Moss Pink
Occasional. Cemeteries, open ledges. Fl. 5(4)-7. Sheffield. West Stockbridge. Native to farther west and south.

HYDROPHYLLACEAE

HYDROPHYLLUM

H. canadense L. **E** Broad Waterleaf
Rare. In rich moist soil, in ravines and base of calcareous ledge. Fl. 6(3). North Adams, Williamstown. Plate 6C.

H. virginianum L. Waterleaf
Common. Somewhat rich to rich woods. Fl. 6(1-4).

BORAGINACEAE

CYNOGLOSSUM

C. boreale Fern. **H** Northern Wild Comfrey
Rare and historical. Rich moist open woods, calcareous ledges. Stockbridge: <u>Hoffmann</u> 1904 (NEBC).

***C. officinale** L. Common Hound's Tongue
Occasional. Pastures, old fields, border of woods. Fl. 5(4)-6(3). Pittsfield, Williamstown. Introduced from Europe.

ECHIUM

***E. vulgare** L. Viper's Bugloss
Common. Dry open waste areas, roadsides. Fl. 6(3)-8. White forms seen occasionally. Introduced from Europe.

HACKELIA

H. virginiana (L.) Johnst. Stickseed
Frequent. Borders of rich woods. Fl. and fr. 8(1-4).

LAPPULA

***L. echinata** Gilib. Stickseed
Uncommon. Waste areas. Fl. 7(3). Stockbridge: <u>Hoffmann</u> 1916 (NEBC). Introduced from Europe.

LITHOSPERMUM

***L. arvense** L. Corn Gromwell
(**Buglossoides arvense** (L.) Johnst.)
Uncommon. Waste areas. Fl. 6(1). Great Barrington: <u>Hoffmann</u> 1916 (NEBC). Introduced from Europe.

***L. officinale** L. European Gromwell
Frequent. Pastures, roadsides, river banks. Fr. 6(2). Introduced from Europe.

MYOSOTIS

M. laxa Lehm. Wild Forget-me-not
Uncommon. Wet margins of marshes, marshy border of lake, lightly shaded. Fl. 6(2)-9. Monterey, Sheffield.

***M. scorpioides** L. Forget-me-not
Common. Marshes, along brooks, wet meadows. Fl. 6(3)-9. Introduced from Europe.

***M. sylvatica** Hoffm. Garden Forget-me-not
Uncommon. Moist rich roadsides. Fl. 5(2-4). Lenox, North Adams, Williamstown. Introduced from Europe. First discovered in the county by the author in 1989.

SYMPHYTUM

***S. officinale** L. Comfrey
Occasional. Moist roadsides. Fl. 6-8. Introduced from Europe.

VERBENACEAE

PHRYMA

P. leptostachya L. Lopseed
Frequent. Open woods, borders of rich woods. Fl. 7(3)-8(1).

VERBENA

***V. bracteata** Lag. & Rodr. Prostrate Vervain
Uncommon. Waste areas. Fl. 7-8. Pittsfield: <u>Hoffmann</u> 1919 (NEBC). Native farther west.

V. hastata L. Blue Vervain
Common. Wet meadows, marshes, border of ponds. Fl. 7(2)-8. White forms are seen occasionally.

V. Xengelmannii Moldenke
(**V. hastata** X **V. urticifolia**)
Occasional. Pond margins, moist soil. Fl. 7-8. Williamstown.

V. simplex Lehm. **E** Narrow-leaved Vervain
Rare. In dry open calcareous ledges, marble quarry. Fl. 7. Sheffield. Plate 6D.

V. urticifolia L. var. **urticifolia** White Vervain
Common. Borders of ponds, ditches, moist waste areas. Fl. 7-8.

LAMIACEAE

ACINOS

***A. arvensis** (Lam.) Dandy Mother-of-thyme
(**Satureja acinos** (L.) Scheele)
Occasional. Dry calcareous fields, ledges. Lanesboro, Sheffield,Williamstown. Introduced from Europe.

AGASTACHE

A. scrophulariifolia (Willd.) Kuntze **H** Purple Giant Hyssop
Rare and historical. In rich soil, roadside thickets, near rivers. Fl. 7(4)-8. Alford, Sheffield. New Marlboro: Hoffmann 1920 (NEBC). It is difficult to determine why this species cannot be relocated.

AJUGA

***A. genevensis** L. Bugleweed
Uncommon. Dry calcareous soil in pastures. Fl. 5(4)-8(2). West Stockbridge, Williamstown. Introduced from Europe.

***A. reptans** L. Bugleweed
Uncommon. Moister situations than preceding, wet margins of brooks. Fl. 5(2). Williamstown. Introduced from Europe.

BLEPHILIA

B. ciliata (L.) Raf. **E** Downy Wood-mint
Rare. Dry calcareous soil in open woods, abandoned quarries, somewhat disturbed areas. Fl. 7(1-3). Sheffield, Stockbridge, West Stockbridge. Plate 6E.

B. hirsuta (Pursh) Benth. var. **hirsuta E** Hairy Wood-mint
Rare. Rich moist to wet lightly shaded soil, deciduous rich wooded slopes, rich wooded alluvial soil. Fl. 7(3)-8(3). Adams, Cheshire, Egremont, New Ashford, Sandisfield, West Stockbridge. Plate 6F.

CLINOPODIUM

***C. vulgare** L. Wild Basil
(**Satureja vulgaris** (L.) Fritsch)
Common. Dry calcareous soil, pastures, fields and borders of woods. Fl. 6(2)-8. Introduced from Europe.

COLLINSONIA

C. canadensis L. Horse Balm
Frequent. Rich soil, dry open woods. Fl. 7(4)-8(3). Mostly in valley region.

GALEOPSIS

***G. bifida** Boenn. Hemp-nettle
Occasional. Waste places, roadsides. Fl. 8(1). Introduced from Eurasia.

***G. tetrahit** L. Hemp-nettle
Common. Waste and disturbed areas. Fl. 7(3). Introduced from Eurasia.

GLECHOMA

***G. hederacea** L. Ground-ivy, Ground-mint
Common. A weed in moist lawns, gardens, roadsides. Fl. 4(2)-7. Introduced from Europe.

HEDEOMA

H. hispidum Pursh Rough Pennyroyal
Uncommon. Dry calcareous soil, on ledges, quarries, pastures. Fl. 6(3)-7. Great Barrington, Sheffield.

H. pulegioides (L.) Pers. American Pennyroyal
Occasional. Dry somewhat calcareous soil, open rocky woods on south-facing slopes, borders of woods, pastures. Fl. 8.

LAMIUM

***L. amplexicaule** L. Dead-nettle
Uncommon. Waste or disturbed areas, lawns. Fl. 6-8. Pittsfield: Hoffmann 1916 (NEBC). Introduced from Europe.

***L. maculatum** L. Spotted Dead-nettle
Uncommon. Waste areas, roadsides. Fl. 7. Stockbridge: DeGoersdorff 1934 (MASS). Introduced from Europe.

***L. purpureum** L. Purple Dead-nettle
Uncommon. Moist disturbed soil. Fl. 4(4). Williamstown. Introduced from Europe. First discovered in the county by the author in 1990.

LEONURUS

***L. cardiaca** L. Motherwort
Common. Waste areas, roadsides, old fields. Fl. 6(2)-8. Introduced from Europe.

LYCOPUS

L. americanus Muhl. ex Bart. Water-horehound
Frequent. Marshes, wet meadows, borders of streams, ponds. Fl. 7(4)-9.

L. uniflorus Michx. Common Water-horehound
Frequent. Wet, often peaty soil, edges of boggy ponds, marshes. Fl. 7(2)-8.

L. virginicus L. Water-horehound
Frequent. Wet soil, marshes, river banks. Fl. 7(4)-8.

MENTHA

M. canadensis L. Hairy Field Mint
Common. Wet soil, wet meadows, marshes, along brooks and rivers. Fl. 7(3)-8.

***M. Xgracilis** Sole Downy Whorled Mint
(**M. arvensis X M. spicata**)
Occasional. Wet soil on river shores. Fl. 7(4)-8(2). Sandisfield.

***M. longifolia** L. Horse-mint
Uncommon. Dry fields. Fl. 9. Alford. Introduced from Europe.

***M. Xpiperita** L. Peppermint
(**M. aquatica** L. X **M. spicata** L.)
Frequent. Moist soil in wet meadows, springs, streams. Fl. 8-9. Introduced from Europe.

***M. Xrotundifolia** (L.) Huds. Roundleaf Mint
(***M. longifolia** X ***M. suaveolens** Ehrh.)
Uncommon. Wet soil along stream in pasture. Fl. 9. Williamstown. Introduced from Europe.

***M. spicata** L. Spearmint
Frequent. Wet meadows, springs, along streams. Fl. 7(4)-8. Introduced from Europe.

MONARDA

***M. didyma** L. Bee Balm
Uncommon. Roadside thickets. Fl. 7(2)-8(1). Sheffield: Churchill 1920 (NEBC). A garden escape, native farther west.

M. fistulosa L. var. **fistulosa** Wild Bergamot
Occasional. Dry calcareous soil, fields, roadsides. Fl. 8. Sheffield: Churchill 1919 (NEBC).

M. fistulosa L. var. **mollis** (L.) Benth. Wild Bergamot
Common. Dry calcareous meadows, outcrops. Fl. 8.

NEPETA

***N. cataria** L. Catnip, Catmint
Frequent. Around old farm sites, on calcareous ledges, waste areas. Fl. 8-9. Introduced from Europe.

ORIGANUM

***O. vulgare** L. Marjoram, Oregano
Frequent. Dry calcareous soil, old fields. Fl. 7(3)-9. Introduced from Europe.

PHYSOSTEGIA

***P. virginiana** (L.) Benth. False Dragonhead, Obedient Plant
Occasional. Roadsides, fields. Great Barrington, Sheffield, West Stockbridge. Native farther west and south.

PRUNELLA

P. vulgaris L. ssp. **lanceolata** (Barton) Hulten
American Self-heal, Heal-all
Frequent. Moist fields, along brooks, trails, waste areas, lawns. Fl. 6(4)-9. Occasional white-flowered plants are seen.

***P. vulgaris** L. ssp. **vulgaris** Eurasian Heal-all
Common. Borders of fields, open woods, pastures. Introduced from Eurasia.

PYCNANTHEMUM

P. muticum (Michx.) Pers. Mountain-mint
Occasional. Dry calcareous soil, old fields. Fl. 8. Great Barrington, Lanesboro, Williamstown.

P. tenuifolium Schrad. Slender Mountain-mint
Frequent. Dry soil , open fields, roadsides, disturbed soil. Fl. 8-9.

P. virginianum (L.) Dur. & Jacks. ex Fern. & Robins.
Mountain-mint
Frequent. Moist calcareous soil, fields, pastures. Fl. 7(3)-8.

SALVIA

***S. nemorosa** L. Woodland Sage
(**S. sylvestris** L.)
Uncommon. Dry sandy field with calcareous outcrops. Fl. 8. Sheffield. Introduced from Europe. First discovered in the county by B. A. Sorrie in 1989.

SCUTELLARIA

S. galericulata L. Marsh Skullcap
(**S. epilobiifolia** Hamilt.)
Common. Borders of marshes, wet meadows. Fl. 7(3)-8.

S. lateriflora L. Mad-dog Skullcap
Common. Wet meadows, wet woods, along small streams. Fl. 7-8.

STACHYS

S. hispida Pursh Smooth Hedge-nettle
Uncommon. Wet soil. Fl. 8. Sheffield. First discovered in the county by B. A. Sorrie in 1989.

***S. palustris** L. var. **palustris**
Woundwort, Marsh Hedge-nettle
Uncommon. Rich moist soil. Marshy borders of small streams, ponds, wet woods. Fl. 7(3). New Ashford. Williamstown. Introduced from Europe.

TEUCRIUM

***T. botrys** L. Cut-leaf Germander
Uncommon. Dry calcareous outcrops in pastures. Fl. 7(3)-8(2). Great Barrington, Sheffield, Williamstown. Introduced from Europe.

T. canadense L. var. **virginicum** (L.) Eat. American Germander
Locally common. Alluvial soil along Housatonic R. Fl. 8-9. Williamstown, Sheffield.

T. canadense L. var. **occidentale** (Gray) McClintock & Epling
Hairy Germander Uncommon. River banks. Fl. 8. Sheffield: Churchill 1920 (NEBC).

THYMUS

***T. pulegioides** L. Wild Thyme
(**T. serpyllum** L. ssp. **chamaedrys** (Fries) Voll.)
Common. Dry calcareous soil, fields, roadsides, waste areas. Fl. 8-9. Introduced from Europe.

TRICHOSTEMA

T. dichotomum L. Blue-curls
Occasional. Dry soil, waste areas, sandy fields, railroads.. Fl. 8.

T. brachiatum L. **E** False Pennyroyal
(**Isanthus brachiatus** (L.) BSP.)
Rare. On exposed calcareous ledges and outcrops, quarry edges. Fl. 8-9. Sheffield, Williamstown.

CALLITRICHACEAE

CALLITRICHE

C. heterophylla Pursh Water-starwort
Common. Margins of rivers in quiet water, pools in marshes. Fl. and fr. 7-8.

C. verna L. Water-starwort
Common. On mud in pools along streams, rivers, marshes, in gravel of rivers. Fl. and fr. 7-8.

PLANTAGINACEAE

PLANTAGO

***P. arenaria** Waldst. & Kit. Flaxseed Plantain
Uncommon. Roadsides, waste areas. Fl. 8(4). Cheshire, Pittsfield. Introduced from Europe. First discovered in the county by B. A. Sorrie in 1985.

***P. aristata** Michx. Bracted Plantain
Uncommon. Dry fields and waste ground. Fl. 8(3). Stockbridge: Hoffmann 1902 (NEBC). Frequent in the southern part of valley, according to Hoffmann (1922). Native farther west.

***P. lanceolata** L. Ribgrass
Common. Fields, roadsides, waste areas. Fl. 5(4)-8. Introduced from Europe.

***P. major** L. Common Plantain, White-man's Foot
Common. Lawns, roadsides, waste areas. Fl. 7-8. Introduced from Europe.

***P. media** L. Hoary Plantain
Uncommon. Mown meadows and lawns. Great Barrington Fairgrounds. Great Barrington, Lenox. From Europe.

P. rugelii Dcne. Redstem or Pale Plantain
Common. Fields, roadsides, waste areas. Fl. 7-8.

***P. virginica** L. Hoary or Pale-seed Plantain
Uncommon. Dry ground. Fl. 7. Sheffield: S. W. Bailey 1948 (NEBC).

OLEACEAE

FORSYTHIA

***F. suspensa** (Thunb.) Vahl. Forsythia
Uncommon. Roadsides., escaped from plantings. Fl. 5(1). Great Barrington: Hoffmann 1919 (NEBC). Introduced from Asia.

FRAXINUS

F. americana L. White Ash
Common. Moist rich woods and moister parts of hardwood forests. Fl. 5.

F. nigra Marsh. Black Ash
Frequent. Forested fens and swamps.

F. pennsylvanica Marsh. Red Ash, Green Ash
(Including varieties)
Frequent. Alluvial soil. Locally frequent along Housatonic R.

LIGUSTRUM

***L. obtusifolium** Sieb. & Zucc. Privet
Uncommon. Roadside thicket. Fl. 6(2). Great Barrington, Stockbridge. Introduced from Asia.

***L. vulgare** L. Privet

Common. Hedgerows, roadsides, calcareous outcrops. Fl. 6(3)-7(1). Introduced from Europe.

SYRINGA

*__S. persica__ L. Persian Lilac
Occasional. Old fields, hedgerows, roadsides. Fl. 6(2-3). Pittsfield, Williamstown. Introduced from Asia.

*__S. vulgaris__ L. Lilac
Occasional. Roadsides, old fields, near old house sites. Fl. 5(3-4). Introduced from Europe.

SCROPHULARIACEAE

AGALINUS

A. paupercula (Gray) Britt. var. **paupercula**
(**Gerardia paupercula** (Gray) Britt.) Small Gerardia
Uncommon. Wet soil, seepage area in powerline cut. Fl. 9(2). Otis. First discovered in the county by B. A. Sorrie and the author in 1985.

A. tenuifolia (Vahl) Raf. Slender Gerardia
(**Gerardia tenuifolia** Vahl)
Uncommon. Open wooded slopes, powerline cut, disturbed pondshore. Fl. 9(4). Great Barrington, Pittsfield, Sheffield, Stockbridge.

AUREOLARIA

A. flava (L.) Farw. var. **flava** Smooth False-foxglove
Occasional. South-facing slopes in open woods with **Quercus** spp. on which **Aureolaria** spp. are parasitic. Fl. 7(4)-8(4). Hancock, Mt. Washington, Sheffield, West Stockbridge, Williamstown.

A. pedicularia (L.) Raf. Fern-leaved False-foxglove
Uncommon. Dry open woods on south-facing slope. Fl. 8(2). Sheffield. Great Barrington: Hoffmann 1911 (NEBC).

A. virginica (L.) Pennell Downy False-foxglove
Uncommon. Dry rich calcareous soil, open woods. Fl. 7(4)-8(2). Great Barrington, Mt. Washington. New Marlboro. Sheffield.

CHAENORRHINUM

*__C. minus__ (L.) Lange Dwarf Snapdragon
Common. Along railroads, in gravel roadsides, old quarries. Fl. 6(4)-7. Introduced from Europe.

CHELONE

C. glabra L. var. **glabra** Turtlehead
Common. Borders of swamps and brooks, wet woods. Fl. 7 (4)-9.

EUPHRASIA

*__E. micrantha__ Reichenb. Eyebright
Uncommon. Open disturbed areas. Fl. 8-9(2). Adams. First collected in Adams in 1944 and on summit of Mt. Greylock in 1952 by A. K. Harrison, but these specimens were first identified as **E. canadensis** Townsend. F. J. Hermann collected and identified **E. micrantha** on Mt. Greylock in 1957. (See Hermann, 1958). This species may be spreading in the adjacent towns of Adams, Lanesboro, North Adams, Williamstown and in Washington. Introduced from Europe.

*__E. nemorosa__ (Pers.) Wallr. Eyebright
(**E. americana** Wettst., **E. canadensis** Townsend)
Frequent. Disturbed moist roadsides, old fields. Fl. 6(4)-9(1). Adams, Becket, Cheshire, Florida, Lanesboro, Savoy. First discovered in the county in Savoy by H. E. Ahles in 1972. Introduced from Europe?

GRATIOLA

G. aurea Muhl. Golden Pert
Uncommon. Submersed in shallow water of acidic pond in vegetative state. On shore on rotted log in flowering state. Fl. 9(1). Savoy. First discovered in the county by the author in 1992.

G. neglecta Torr. Clammy Hedge-hyssop
Occasional. On mud in abandoned beaver ponds, ditches, marshes, along rivers, disturbed areas. Fl. 7(2)-8.

LINARIA

L. canadensis (L.) Dumort. Old-field Toadflax
Occasional. Dry sandy open soil. Fl. 6(3)-9. Florida, Great Barrington, Sheffield, Williamstown.

*__L. vulgaris__ Mill. Butter and Eggs
Common. Roadsides, dry fields, waste areas. Fl. 6(3)-8. Introduced from Europe.

LINDERNIA

L. dubia (L.) Pennell var. **dubia** False Pimpernel
Frequent. In alluvial soil or mud on river banks, mud bars in streams, bare soil in wet meadows, moist cultivated fields. Fl. 7(3)-8.

MELAMPYRUM

M. lineare Desr. Cow-wheat
(Including vars. **americana** (Michx.) Beauv., **latifolia** Bart.)
Common. Sterile dry soil in open acidic woods. Fl. 6(2)-8.

MIMULUS

M. ringens L. Monkeyflower
Common. Wet meadows, marshes, borders of streams. Fl. 6(4)-8.

PEDICULARIS

P. canadensis L. Lousewort, Wood Betony
Common. Dry rich soil in open woods, alluvial sandy soil. Fl. 5(2)-6(1).

PENSTEMON

*__P. calycosus__ Sm. Beard-tongue
Uncommon. Calcareous soil, disturbed soil of roadsides and old fields. Fl. 6(4). Richmond, Sheffield. Native farther west and south. First discovered in the county by B. A. Sorrie in 1988.

*__P. digitalis__ Nutt. White Beard-tongue
Occasional. Moist calcareous soil in meadows, roadside, clearings. Fl. 6(4)-7(2). Egremont, Sheffield, Williamstown. Native farther west and south.

P. hirsutus (L.) Willd. **E** Hairy Beard-tongue
Rare. On dry exposed calcareous ledges. Fl. 6(3)-7(1). Sheffield. Plate 7A.

*__P. pallidus__ Sm. Pale Beard-tongue
Uncommon. Disturbed ground, roadside. Fl. 6(2). Becket: Sorrie 3831 1987 (NEBC). Native farther west and south. First discovered in the county by B. A. Sorrie in 1987.

RHINANTHUS

*__R. minor__ L. Yellow-rattle
(**R. crista-galli** L.)
Frequent. Disturbed soil, roadsides, moist fields. Fl. 6(2)-7. Native farther north or from Europe, now spreading rapidly throughout the county. Not listed by Hoffmann.

SCROPHULARIA

S. lanceolata Pursh Hare-figwort
Occasional. Roadsides, fields, farmyard. Fl. 6(4)-8(1).

S. marilandica L. Carpenter's-square, Figwort
Occasional. Rich soil below calcareous ledges, rich alluvial soil. Fl. 6(4)-9(2).

VERBASCUM

*__V. blattaria__ L. Moth-mullein
Occasional. Fields. Fl. 6(3)-8. Egremont, Williamstown. Introduced from Europe.

*__V. thapsus__ L. Mullein
Common. Waste areas, roadsides, logging clearings in woods. Fl. 6(4)-9. Introduced from Europe.

VERONICA

V. americana (Raf.) Schwein. ex Benth. American Speedwell
Common. Along small brooks, springs, borders of marshes, wet meadows. Fl. 6(4)-7(4).

V. anagallis-aquatica L. Water Speedwell
Occasional. Alluvial soil, banks of rivers, springs. Fl. 6(3)-9(2). Egremont. Great Barrington, Lee. Native status uncertain.

*__V. arvensis__ L. Corn Speedwell
Common. Lawns, gardens, waste areas. Fl. 5(3)-7(3). Introduced from Europe.

*__V. austriaca__ L. ssp. **teucrium** (L.) Webb Broadleaf Speedwell
(**V. latifolia** L.)
Uncommon. Roadsides near old house sites. Fl. 6(3-4). New Marlboro, Williamstown. Introduced from Europe.

V. catenata Pennell **E** Water Speedwell
(**V. comosa** Richter, **V. connata** of authors, not Raf.)
Rare. On mud in intermittent stream flowing through wooded swamp, border of pond at entrance of small stream from calcareous marsh. Fl. 8(2)-7(1). Sheffield. Plate 7B.

*__V. chamaedrys__ L. Bird's-eye Speedwell
Frequent. Rich soil on roadsides, pastures. Fl. 5(2)-9(4). Introduced from Europe.

*__V. officinalis__ L. Common Speedwell
Common. Dry open fields and woods, disturbed areas. Fl. 6(1)-7. Both native and introduced from Europe (Fernald, 1950).

V. peregrina L. ssp. **peregrina** Neckweed
Common. Weed of lawns and gardens. Fl. 6(1)-7(1).

*__V. persica__ Poir. Bird's Eye Speedwell
Common. Moist lawns. Fl. 4(3)-5. Introduced from Eurasia.

V. scutellata L. Marsh Speedwell
Frequent. On mud among marsh vegetation, stream banks, wet woods. Fl. 7(3)-8(3).

*__V. serpyllifolia__ L. ssp. **serpyllifolia** Thyme-leaf Speedwell
Common. Weed of lawns and gardens. Fl. 5(2)-7(2). Introduced from Europe.

VERONICASTRUM

V. virginicum (L.) Farw. **SC** Culver's Root
Rare. In rich alluvial soil, wet calcareous meadows, shrub swamps. Often at edge of these habitats. Fl. 8(1)-9(1). Egremont, Great Barrington, Monterey, Pittsfield, Stockbridge, Sheffield.

OROBANCHACEAE

CONOPHOLIS

C. americana (L.f.) Wallr. Squawroot
Occasional. Rich soil, in rocky woods. Fl. 6(3). Adams, Hancock, Pittsfield, Richmond. Plate 7C.

EPIFAGUS

E. virginiana (L.) Bartr. Beech-drops
Common. Under American beech and parasitic on its roots. Fl. 8. A pure yellow form was collected on Mt. Greylock, referred to as f. **pallida**.

OROBANCHE

O. uniflora L. One-flowered Cancer-root
Occasional. Rich rocky woods. Fl. 6(3). Adams, Hancock, Pittsfield, Richmond.

BIGNONIACEAE

CATALPA

***C. speciosa** (Warder ex Barney) Engelm. Catalpa
Uncommon. Alluvial soil on river banks and edges of flood-plain marshes. Fl. 7(1). North Adams, Sheffield. Escaped from cultivation and may be spreading in the above habitats (See Burk & Prabhu, 1988). Native farther south. Not listed by Hoffmann.

LENTIBULARIACEAE

UTRICULARIA

U. cornuta Michx. Horned Bladderwort
Occasional. In bare peat at edge of pools in Sphagnum bogs and boggy ponds. Fl. 7(3)-9(2). Mt. Washington, Otis, Savoy.

U. geminiscapa Benj. Bladderwort
Uncommon. Acidic cold ponds. Fl. 7(3). Mt. Washington, Otis.

U. gibba L. Humped Bladderwort
Uncommon. On open peat at edge of pools in Sphagnum bog mat or peaty edge of marshes. Vegetative branches floating. Fl. 8(3). Egremont, Sandisfield, Savoy.

U. intermedia Hayne Milfoil Bladderwort
Occasional. In channels in open graminoid fens and calcareous marshes. Fl. 7(2). Egremont, Richmond, Sheffield, Stockbridge.

U. minor L. Lesser Bladderwort
Uncommon. Pools in marshes and bogs. Fl. 6(3-4). Egremont, Monterey, Pittsfield, Stockbridge.

U. purpurea Walt. Purple Bladderwort
Frequent. Acidic boggy ponds. Fl. 7(4)-8. Cheshire, Clarksburg, Mt. Washington, Otis, Sandisfield. Hoffmann (1922) did not list this species, but it was collected in 1895 in Mt. Washington by R. R. Tatnall (GH). Apparently, it has become much more abundant.

U. radiata Small Small Floating Bladderwort
Frequent. Acidic boggy ponds. Fl. 8. Mostly on the Berkshire Plateau. Clarksburg, Becket, Great Barrington, Otis, Savoy. Sandisfield. Not listed by Hoffmann (1922). Possibly introduced and spread by increased boat use. More habitat has been created by the conversion of bogs and marshes into ponds, particularly on Berkshire Plateau.

U. resupinata B. Greene Bladderwort
Uncommon. Margin of acidic pond. Fl. 7(2). Mt. Washington: Tatnall 1895 (GH). Habitat remains unchanged and there is a possibility it may be relocated.

U. vulgaris L. Common Bladderwort
Common. Pools in marshes, in ponds and lakes, both acidic and alkaline. Fl. 7-8.

CAMPANULACEAE

CAMPANULA

C. aparinoides Pursh Marsh Bellflower
(**C. uliginosa** Rydb.)
Frequent. Wet soil in marshes, reclining on marsh vegetation. Fl. 7(3)-9(1).

***C. rapunculoides** L. Creeping Bellflower
Common. Disturbed and cultivated areas, edge of woods. Fl. 7(3). Introduced from Europe.

C. rotundifolia L. Harebell
Occasional. Dry open calcareous or serpentine ledges, shore of Deerfield River. Occasional on schist ledges. Fl. 7(1)-9. Most common in Sheffield, New Marlboro. Also Florida, Mt. Washington, New Ashford.

LOBELIA

L. cardinalis L. Cardinal Flower
Frequent. Borders of rivers and streams. Fl. 7(3)-8.

L. dortmanna L. Water Lobelia
Occasional. Shallow water in very acidic ponds with rocky or gravel bottom. Fl. 7(1)-8. Becket, Otis, Savoy.

L. inflata L. Indian-tobacco
Common. Dry sterile soil, open woods, fields, roadsides. Fl. 8.

L. kalmii L. Brook or Kalm's Lobelia
Occasional. Wet calcareous soil, fens, wet meadows and pastures, calcareous seeps and brook margins. Fl. 8-9.

L. siphilitica L. **T** Great Blue Lobelia
Rare. In wet calcareous soil, meadows, pastures, river banks. Fl. 8(4)-9. Mt. Washington, Pittsfield, Sheffield, Stockbridge.

L. spicata L. Pale-blue Lobelia
Frequent. In calcareous meadows. Fl. 7(1)-9(2).

TRIODANIS

T. perfoliata (L.) Nieuwl. Venus' Looking-glass
(**Specularia perfoliata** (L.) A. DC.)
Occasional. Dry soil, hillsides and open rocky woods on south-facing slopes. Fl. 6(4). Monterey, Mt. Washington, Sheffield, Williamstown. Hoffmann lists it as "frequent in the southern part of valley."

RUBIACEAE

CEPHALANTHUS

C. occidentalis L. Buttonbush
Frequent. Shrub swamp, borders of marshes, ponds, streams. Fl. 8.

GALIUM

G. aparine L. Cleavers, Goosegrass
Common. Climbing on vegetation in borders of moist woods. Fl. 6-7. Native and introduced from Eurasia.

G. asprellum Michx. Rough Bedstraw
Common. Climbing on vegetation in moist thickets. Fl. 7-8.

G. boreale L. **E** Northern Bedstraw
Rare. Calcareous soil, wet meadows and pastures. Fl. 7(2). Adams, Lenox, Sheffield.

G. circaezans Michx. var. **hypomalacum** Fern. Wild Licorice
Frequent. Dry or moist rich woods. Fl. 6(3)-7(1).

G. labradoricum (Wieg.) Wieg. **SC** Labrador Bedstraw
Rare. In fens and calcareous marshes, usually twining around shrubby vegetation on mossy hummocks in fens. Fl. 5(4)-7(1). Alford, Great Barrington, Lee, Lenox, Richmond, Sheffield, Stockbridge.

G. lanceolatum Torr. Wild Licorice
Frequent. Moist rich woods. Fl. 6(2)-7(1).

***G. mollugo** L. White Bedstraw, Wild Madder
Common. Abundant in fields and meadows. Fl. 6(2)-8(2). Introduced from Europe.

G. obtusum Bigel. Large Marsh Bedstraw
Frequent. Open marshes. Fl. 7(1).

G. palustre L. Marsh Bedstraw
Common. Wet soil, marshes, wet meadows. Fl. 6(2)-7(1).

G. pilosum Ait. Hairy Bedstraw
Uncommon. Dry calcareous soil, south-facing slope in open woods. Fl. 8(1). West Stockbridge, New Ashford, Sheffield.

G. tinctorium L. Clayton's Bedstraw
Occasional. Cool bogs and fens, swamps. Fl. 7(2-3).

G. trifidum L. ssp. **trifidum** Small Bedstraw
Frequent. Marshes, in shallow water in ponds, vegetative plant sometimes submersed. Fl. 7(4)-8(4).

G. triflorum Michx. Sweet-scented Bedstraw
Common. Rich moist woods. Fl. 6(4)-8(2).

***G. verum** L. Our-lady's Bedstraw
Common. Fields and roadsides. Fl. 6(3)-7(3). Common in southern towns, has been spreading north.in the last few years. Introduced from Europe.

HEDYOTIS

H. caerulea (L.) Hook. Bluets, Quaker Ladies
(**Houstonia caerulea** L.)
Common. Fields, pastures, roadsides, clearings, disturbed areas. Fl. 4(4)-7.

H. longifolia (Gaertn.) Hook. **T** Longleaf Bluets
(**Houstonia longifolia** Gaertn.)
Rare. On open rocky ledges, in crevices and old quarry cuts in full sun. Fl. 6(3)-7. West Stockbridge.

MITCHELLA

M. repens L. Twinberry, Partridge-berry
Common. Moist deciduous woods. Fl. 7(1-3).

CAPRIFOLIACEAE

DIERVILLA

D. lonicera Mill. Bush Honeysuckle
Frequent. On acidic rocky cliffs, banks, open or lightly shaded. Fl. 6(2-4).

LINNAEA

L. borealis L. Twinflower
Uncommon. Wet cool spruce-fir forest, cool mossy woods. Fl. 6(4)-7(3). Mt. Washington, New Marlboro, Peru., Savoy, Washington. Plate 7D.

LONICERA

*__L. bella__ Zabel Honeysuckle
(**L. morrowii** X **L. tatarica**)
Common. Roadsides, edges of woods, fields. Fl. 5(4). Introduced from Asia.

L. canadensis Bartr. American Fly Honeysuckle
Frequent. Moist soil, cool open woods. Fl. 4(4)-6(1).

L. dioica L. Glaucous or Limber Honeysuckle
Frequent. Calcareous soil, borders of marshes and seeps, rich open woods, open ledges. Fl. 5(3).

L. hirsuta Eat. var. **hirsuta E** Hairy Honeysuckle
Rare. On open calcareous ledges and open rich woods. Fl. 6(2-3). Williamstown, Lee. Type locality is in Williamstown, where Amos Eaton's student Emory Washburn discovered the plant in 1817. Plate 7E.

*__L. morrowii__ Gray Morrow's Honeysuckle
Common. Abundant on roadsides, open woods and ledges, floodplains, crowding out native vegetation. Fl. 5(4)-6(2). Introduced from Eurasia. The introduced honeysuckles have become serious pests since 1922, particularly **L. morrowii**. Not noted by Hoffmann.

*__L. sempervirens__ L. Trumpet Honeysuckle
Uncommon. An escape on rocky hillside. Fl.7(1). Great Barrington: Hoffmann 1911 (NEBC). Native farther east and south.

*__L. tatarica__ L. Tatarian Honeysuckle
Uncommon. Alluvial soil along river, roadside. Fl. 5(4). New Marlboro, Sheffield. Introduced from Eurasia.

L. villosa Michx. Mountain Fly Honeysuckle
Occasional. Calcareous soil, in marshes, fens, wet meadows, seeps. Fl. 5(3-4).

*__L. xylosteum__ L. European Fly Honeysuckle
Uncommon. Hedgerows, open woods. Fl. 5(4). Lenox, Sheffield, Stockbridge. Introduced from Eurasia.

SAMBUCUS

S. canadensis L. Elderberry
Common. Moist soil, hedgerows, borders of ponds and rivers. Fl. 6(4)-7(3).

S. racemosa L. ssp. **pubens** (Michx.) House
(**S. pubens** Michx.) Red-berried Elderberry
Frequent. Cool woods, along roads and trails. Fl. 5(3).

SYMPHORICARPOS

S. albus (L.) Blake var. **albus H** Snowberry
Rare and historical. On dry calcareous hill. Fl. 6(3). Sheffield: Hoffmann 1902 (NEBC).

*__S. albus__ (L.) Blake var. **laevigatus** (Fern.) Blake
Western Snowberry
Occasional. Along roads, escaped from cultivation. Fl. 6(2). Introduced from Pacific Coast.

TRIOSTEUM

T. aurantiacum Bickn. Wild Coffee, Horse-gentian
Occasional. Dry or moist calcareous soil, open woods. Fl. 6(3). Richmond and southward in the valley region.

VIBURNUM

V. acerifolium L. Maple-leaf Viburnum
Common. Most dry or moist acidic or calcareous woods. Fl. 6(1-4).

V. cassinoides L. Wild Raisin, Withe-rod
Frequent. Moist soil in clearings, roadsides, old fields, marshes. Fl. 6(2-4).

*__V. lantana__ L. Wayfaring Tree
Uncommon. Moist rich woods. Fl. 5(3). Stockbridge. Introduced from Europe.

V. lantanoides Michx. Hobblebush
(**V. alnifolium** Marsh.)
Frequent. Cool moist woods. Fl. 5(1-4). More common in moist mature forests at higher elevations.

V. lentago L. Nanny-berry
Common. Wet soil on pond and marsh margins, swamps, roadsides. Fl. 5(4)-6(4).

*__V. opulus__ L. Guelder-rose

Occasional. Margins of marshes, wet soil. Fl. 6(2). Introduced from Europe.

V. rafinesquianum Schultes **T** Downy Arrowwood
Rare. In open dry rocky somewhat calcareous woods on ridgetops or south-facing slopes. Fl. 6(1-2). West Stockbridge, Mt. Washington.

V. recognitum Fern. Arrowwood
Common. Margins of marshes, shrub swamps, moist roadsides. Fl. 6(2-3).

V. trilobum Marsh. Highbush Cranberry
Frequent. Marshes, wooded swamps, moist roadsides. Fl. 5(4)-6(2).

VALERIANACEAE

VALERIANA

***V. officinalis** L. Common Valerian
Frequent. Wet meadows, roadsides. Fl. 6(3)-7(3). Introduced from Europe.

VALERIANELLA

***V. umbilicata** (Sullivant) Wood Corn-salad
Uncommon. Stockbridge: Shear 1891 (GH). Introduced from farther west.

DIPSACACEAE

DIPSACUS

***D. sylvestris** Huds. Teasel
Uncommon. Moist calcareous fields. Fl. 8(1). Sheffield. Introduced from Europe.

KNAUTIA

***K. arvensis** (L.) Coulter Blue-buttons
Uncommon. Pasture, moist roadside. Fl. 6. Lanesboro, Cheshire. Introduced from Europe. First discovered in the county by H. E. Ahles in 1972.

SUCCISELLA

***S. inflexa** (Kluk) Beck Devil's Bit
Occasional. Alluvial soil along Housatonic River. Fl. 9-10. Lee, Pittsfield. Introduced from Europe.

ASTERACEAE

ACHILLEA

***A. millefolium** L. var. **millefolium** Yarrow
Common. Fields, roadsides, waste areas. Fl. 6(4)-8. Introduced from Europe.

***A. ptarmica** L. Sneezeweed
Uncommon. Wet soil along streams. Fl. 7(4)-8(3). Florida. A garden escape. Introduced from Eurasia. First discovered in the county by H. E. Ahles in 1979.

AMBROSIA

A. artemisiifolia L. Ragweed
Common. Roadsides, waste places. Fl. 8(2)-9.

A. trifida L. Giant Ragweed
Occasional. Alluvial soil on Hoosic River floodplain and in a few farmyards as if introduced with seed. Fl. 8(1-3). Stockbridge, Williamstown.

ANAPHALIS

A. margaritacea (L.) Benth. & Hook.f. ex Clarke
(Including subspecies and varieties) Pearly Everlasting
Common. Fields, clearings, dry woods. Fl. 7(3)-9(2).

ANTENNARIA

A. neglecta Greene var. **attenuata** (Fern.) Cronq. Pussytoes
(**A. howellii** Greene ssp. **neodioica** (Greene) Bayer)
Common. Dry fields, open dry woods. Fl. 5(4)-6(1).

A. neglecta Greene var. **randii** (Fern.) Cronq. Pussytoes
(**A. howellii** Greene ssp. **canadensis** (Greene) Bayer)
Common. Fields, pastures and open woods. Fl. 5(1-4).

A. plantaginifolia (L.) Richards var. **plantaginifolia**
(**A. petiolata** Fern.) Plantain-leaved Pussytoes
Frequent. Rocky acidic soil, open woods. Fl. 5(1-3).

A. plantaginifolia (L.) Richards. var. **ambigens** (Greene) Cronq. Plantain-leaved Pussytoes
(**A. parlini** Fern. ssp. **fallax** (Greene) Bayer & Stebbins)
Common. Dry banks, open woods, roadsides. Fl. 5(4)-6(1).

A. plantaginifolia (L.) Richards. var. **arnoglossa** (Greene) Cronq. Smooth Pussytoes
(**A. parlinii** Fern. ssp. **parlini**)
Frequent. In more calcareous soil than other varieties. Fl. 5(3).

ANTHEMIS

***A. cotula** L. Mayweed, Stinking Chamomile
Common. Pastures, on roadsides. Fl. 7(1). Introduced from Europe.

ARCTIUM

***A. lappa** L. Great Burdock
Frequent. Moist soil along roads, in disturbed areas. Fl. 7(4). Introduced from Europe.

***A. minus** (Hill) Bernh. Common Burdock
Common. Waste areas, old fields, pastures, roadsides. Fl. 8(2). Introduced from Europe.

***A. nemorosum** Lej. & Court. Burdock
Uncommon. Roadsides. Fl. 9(3-4). Lee, Stockbridge. Introduced from Europe.

***A. tomentosum** Mill. Burdock
Uncommon. Roadsides. Fl. 9. Lee, Stockbridge.

ARTEMISIA

***A. absinthium** L. Wormwood
Occasional. Roadsides and waste areas. Fl. 7(4). Becket, New Marlboro. Introduced from Europe.

***A. biennis** Willd. Sage-weed
Occasional. Sandy soil and waste areas. Fl. 9(3-4). Pittsfield, Richmond. Native farther west.

***A. ludoviciana** Nutt. Western Mugwort
Uncommon. Waste areas. Fl. 9(1). Lee: Hoffmann 1920 (NEBC). Native farther west.

***A. vulgaris** L. Mugwort
Common. Roadsides, waste areas. Fl. 8(3)-9(1). Introduced from Europe.

ASTER

The nomenclature of this genus generally follows Gleason's (1952) treatment and Jones (1978, 1980)

A. acuminatus Michx. Whorled Wood Aster
Common. Open woods. Fl. 7(4)-9(1).

A. Xamethystinus Nutt.
(**A. ericoides** X **A. novae-angliae**)
Uncommon. Gravel pit, pasture where it is growing between both parents. Fl. 9(3). North Adams, Williamstown.

A. cordifolius L. Heart-leaved Aster
(Including varieties)
Common. Clearings, along trails and roads. Fl. 8(4)-9.

A. cordifolius X **A. lateriflorus**
Uncommon. Sheffield: Hoffmann 1912 (NEBC).

A. divaricatus L. White Wood Aster
Common. Open woods. Fl. 8(3)-9.

A. ericoides L. White Wreath Aster
Common. Dry open fields, roadsides. Fl. 9.

A. laevis L. Smooth Blue Aster
Frequent. Dry soil, open disturbed sites. Fl. 9.

A. lanceolatus Willd. ssp. **simplex** (Willd) A. G. Jones
Tall White Aster
Common. Moist soil, border of marshes, roadsides. Fl. 9(4)-10(2).

A. lateriflorus (L.) Britt. var. **lateriflorus** Calico Aster
(Including **A. vimineus** Lam.)
Common. Clearings, borders of woods, fields. Fl. 8(3)-9. This taxon hybridizes with numerous other **Aster** spp. Those that probably occur in the county are: **A. lateriflorus** X **A. puniceus**; **A. lateriflorus** X **A. simplex**; **A. lateriflorus** X **A. undulatus**. Specimens of such possible hybrids may be found at NEBC.

A. linariifolius L. Stiff Aster
Occasional. Dry sandy soil, open areas along roads, in sand-plain along roads. Fl. 9. Great Barrington, New Marlboro, Sheffield.

A. macrophyllus L. Large-leaf Aster
(Including var. **macrophyllus**, var. **ianthinus**)
Common. Rich open woods, borders of woods. Fl. 8.

A. novae-angliae L. New England Aster
Common. Moist to wet soil, fields, hedgerows, borders of woods. Fl. 9. Rose-colored forms occur.

A. novi-belgii L. var. **novi-belgii** New York Aster
Occasional. Borders of marshes. Fl. 9-10(1). Mostly in southern third of county. Fl. 9.

A. novi-belgii L. var. **johannensis** (Fernald) A. G. Jones
St. John Aster
Uncommon. On shore of Deerfield River. Fl. 7(4)-9(1). Florida.

A. paternus Cronq. White-topped Aster
(**Sericocarpus asteroides** (L.) BSP.)
Occasional. Dry soil, open south-facing slopes in open woods. Fl. 7(4)-8(4). Great Barrington, Sheffield, West Stockbridge. Mt. Washington.

A. pilosus Willd. var. **pilosus** Heath Aster
Uncommon. Meadows and roadsides. Fl. 9. Stockbridge: Hoffmann 1912 (NEBC).

A. pilosus Willd. var. **demotus** Blake Heath Aster
Common. Dry fields, roadsides, waste areas. Fl 8(4)-9.

A. pilosus Willd. var. **pringlei** (Gray) Blake
(**A. pringlei** (Gray) Britt.) Pringle's Aster
Uncommon. Dry calcareous meadows. Fl. 9. Stockbridge: Hoffmann 1919 (NEBC).

A. prenanthoides Muhl. **SC** Crooked-stem Aster
Rare. Rich alluvial soil along rivers in floodplain forest. Rarely in moist rich soil in upland. Most common along Green Rivers in Great Barrington and Williamstown. Fl. 8(2)-9. Egremont, Great Barrington, Hancock, Lenox,Williamstown. Isolated populations in Washington and Savoy.

A. puniceus L. var. **puniceus** Purple-stemmed Aster
Common. Marshes, borders of streams and rivers, open wooded swamps. Fl. 9. Hybrids with **A. simplex** have been collected (NEBC).

A. puniceus L. var. **firmus** (Nees) A. G. Jones
White Swamp Aster
Occasional. Wet meadows, riverbanks. Most common in southern part of county. Fl. 9(3-4).

A. schreberi Nees Schreber's Aster
Common. Moist rich soil on river banks. Fl. 7(3)-8.

A. umbellatus Mill. Flat-topped Aster
Common. Borders of marshes, stream banks, moist roadsides. Fl. 7(4)-8. Occasional in southern half of county, common in northern half and on Berkshire Plateau.

A. undulatus L. Wavy-leaved Aster
Frequent. Dry open woods and sandy roadsides. Fl. 9. Most frequent in southern half of county.

BELLIS

***B. perennis** L. English Daisy
Occasional. Lawns. Fl. 5(1). Stockbridge, Williamstown. Introduced from Europe.

BIDENS

B. cernua L. Bur-marigold
(Including varieties)
Common. Muddy open margins of marshes, streams. Fl. 8(4)-9.

B. connata Muhl. ex Willd. Swamp Beggar-ticks
Occasional. River banks, wet muddy shaded margins of ponds, marshes. Fl. 8(4)-9. Egremont, New Marlboro, Sheffield. Included in **B. tripartita** by some authors (Gleason, 1952).

B. frondosa L. Beggar-ticks
Common. Wet soil on roadsides, stream banks, wet pastures. Fl. 8(4)-9.

***B. tripartita** L. Beggar-ticks
(**B. comosa** (Gray) Wieg.)
Occasional. Disturbed wet soil, seepage areas, mud along rivers. Fl. 9(4)-10(3). Stockbridge, Williamstown.

B. vulgata Greene Beggar-ticks
Frequent. Disturbed areas, clearings, roadsides. Fl. 8(3)-9.

CENTAUREA

***C. cyanus** L. Bachelor's Buttons
Uncommon. Waste ground, a garden escape. Fl. 7(4). Lee: Hoffmann 1920 (NEBC). Introduced from Europe.

***C. jacea** L. Brown Knapweed
Occasional. Moist rich soil, in fields. Fl. 7(3)-9. Richmond.. Introduced from Europe.

***C. maculosa** Lam. Spotted Knapweed
Common. Abundant on dry roadsides, waste areas, dry fields. An invasive weed in some natural dry habitats. Fl. 7(4)-9. Introduced from Europe.

***C. nigra** L. Black Knapweed
Occasional. Fields. Fl. 7(3)-8. Sandisfield, Williamstown. Introduced from Europe.

CICHORIUM

***C. intybus** L. Chicory, Cornflower
Common. Fields, roadsides, disturbed areas. Fl. 7(2)-9. Introduced from Europe.

CIRSIUM

***C. arvense** (L.) Scop. var. **arvense** (L.) Scop. Canada Thistle
Common. Pastures, roadsides, waste areas. Fl. 7(3)-9. An obnoxious weed. Introduced from Europe.

***C. arvense** (L.) Scop. var. **mite** Wimmer & Grab.
Occasional. Moist clearing, growing with the preceding. Mt. Washington. Introduced from Europe.

C. discolor (Muhl. ex Willd.) Spreng. Field Thistle
Frequent. Dry fields, roadsides. Fl. 8(4)-9.

C. muticum Michx. Swamp Thistle
Occasional. Shrub fens and borders of forested fens. Fl. 8(2-4).

C. pumilum (Nutt.) Spreng. Pasture Thistle
Frequent. Dry old fields and pastures. Fl. 7(4)-9(1).

***C. vulgare** (Savi) Tenore Bull Thistle

Common. Old fields, pastures, waste areas. Fl. 8-9. Introduced from Eurasia.

CONYZA

C. canadensis (L.) Cronq. var. **canadensis** Horseweed
(**Erigeron canadensis** L.)
Common. Newly-disturbed areas, waste areas, abandoned fields. Fl. 7(3)-9.

CREPIS

*__C. capillaris__ (L.) Wallr. Hawk's-beard
Occasional. Disturbed soil, waste areas, fields. 7(4)-8. Stockbridge. Introduced from Europe.

ERECHTITES

E. hieracifolia (L.) Raf. ex DC. Fireweed, Pilewort
Frequent. Burned areas, logging clearings. Fl. 8(4).

ERIGERON

E. annuus (L.) Pers. Daisy Fleabane
Common. Fields, clearings, roadsides, waste areas. Fl. 6(3)-9.

E. philadelphicus L. Pink Fleabane
Frequent. Moist fields and wet meadows. Fl. 6(1-4).

E. pulchellus Michx. Robin's-plantain
Common. More often rich soil. Moist meadows, borders of woods, roadsides. Fl. 6.

E. strigosus Muhl. ex Willd. var. **strigosus** Daisy Fleabane
Common. Dry fields. Fl. 6(2)-8(3).

EUPATORIUM

E. fistulosum Barratt ex Hook. Joe-pye Weed
Uncommon. Moist rich floodplain woods. Fl. 8. Great Barrington, Egremont.

E. maculatum L. var. **maculatum** Spotted Joe-pye Weed
Common. Wet fields and marshes. Fl. 7(4)-9(1).

E. perfoliatum L. Boneset
Common, in wet fields, marshes, roadsides. Fl. 7(4)-9.

E. purpureum L. Sweet Joe-pye Weed
Occasional. Open rich moist woods and borders. Fl. 7(4)-8(3).

E. rugosum Houtt. White Snakeroot
(**Ageratina altissima** (L.) King & H.E. Robins. var. **altissima**)
Common. Rich soil, clearings in woods, borders, open woods. Fl. 9.

E. sessilifolium L. Upland Boneset
Uncommon. Dry calcareous soil, open woods. New Marlboro: Hoffmann 1912 (NEBC).

EUTHAMIA

E. graminifolia (L.) Nutt. ex Cass. var. **graminifolia** Grass-leaved Goldenrod
(**Solidago graminifolia** (L.) Salisb.)
Common. Dry or moist fields and roadsides. Fl. 8(3)-9.

GALINSOGA

*__G. parviflora__ Cav. Quickweed
Frequent. Waste areas, gardens. Fl. 8-10. Introduced from South America.

*__G. quadriradiata__ Ruiz & Pavon Quickweed
(**G. ciliata** (Raf.) Blake)
Common. Abundant weed in gardens, waste areas. Fl. 6(4)-10.

GNAPHALIUM

G. obtusifolium L. Common Everlasting
Common. Dry soil, fields, disturbed areas, waste areas. Fl. 8(3)-9.

G. uliginosum L. Low Cudweed
Common. Alluvial soil on riverbanks, dry soil in open waste areas, roadsides. Fl.8(3)-9.

G. viscosum H. B. K. Cudweed
(**G. macounii** Greene)
Uncommon. Recently disturbed open soil such as abandoned cornfields, roadsides. Fl. 8(2)-10(1).

HELENIUM

H. autumnale L. Sneezeweed
Occasional. Alluvial soil, along streams and rivers. Fl. 8(4)-9. From Lee and Stockbridge south in central valley region.

*__H. flexuosum__ Raf. Sneezeweed
(**H. nudiflorum** Nutt.)
Uncommon. Old fields. Fl. 7(3)-8. Great Barrington: Hoffmann 1920 (NEBC). B. A. Sorrie noted it in Otis, 1984 along Mass. Turnpike (pers. comm.). Native farther south.

HELIANTHUS

*__H. annuus__ L. Sunflower
Occasional. Roadsides. Self-seeding and spread by birds. Fl. 9. No specimen but noted by the author. Introduced from farther west.

H. decapetalus L. Thin-leaf Sunflower
Common. Moist borders of woods and clearings, roadsides. Fl. 8(2)-9(2).

H. divaricatus L. Woodland Sunflower
Occasional. Dry open rocky woods with some calcareous influence. Fl. 8(1-3). Great Barrington, New Marlboro, Sheffield.

***H. mollis** Lam. Ashy Sunflower
Uncommon. Dry soil near railroad tracks. Fl. 8(3-4). Williamstown. Native farther west. First discovered in the county by the author in 1985.

H. strumosus L. Harsh Sunflower
Occasional. Dry open rocky soil, borders of woods, ridges. Fl. 8(2-4). New Marlboro, Sheffield, Williamstown.

***H. tuberosus** L. Jerusalem Artichoke
Occasional. Alluvial soil or silt-gravel bars along rivers, roadsides. Fl. 9. Introduced from farther west.

HELIOPSIS

***H. helianthoides** (L.) Sweet var. **helianthoides** Ox-eye
Uncommon. Fields, waste areas. Fl 7(4). Stockbridge: Hoffmann 1916 (NEBC). Introduced from farther west.

***H. helianthoides** (L.) Sweet var. **scabra** (Dunal) Fisch.
Occasional. Roadsides, along railroads, edges of fields. Sheffield, Stockbridge, Williamstown. Fl. 7(3)-9(3). Introduced from farther west.

HIERACIUM

***H. aurantiacum** L. Devil's Paintbrush
Common. Fields. Fl. 6(4)-9(3). Introduced from Europe.

***H. caespitosum** Dumort King-devil
(**H. pratense** Tausch)
Common. Dry open soil, on roadsides, fields, disturbed areas. Fl. 6(2)-7(4). Introduced from Europe.

H. canadense Michx. var. **fasciculatum** (Pursh) Fern.
Canada Hawkweed
Uncommon. Pond shore, rocky shores of Deerfield R., open old fields. Fl. 8(3)-9(2). Florida, New Ashford, West Stockbridge. Hoffmann (1922) noted it was common; it does not appear to be so at present.

***H. floribundum** Wimmer & Grab. King-devil
Frequent. Open dry disturbed soil. Fl. 6(2)-7. Introduced from Europe.

H. paniculatum L. Panicled Hawkweed
Common. Open somewhat rich woods. Fl. 7(2)-8.

***H. pilosella** L Mouse-ear Hawkweed
Common. Abundant in pastures, often calcareous. Fl. 6(1-4). Introduced from Europe.

***H. piloselloides** Vill. King-devil
(**H. florentinum** All.)
Common. Gravelly waste areas. 6(2)-7(30). Introduced from Europe.

H. scabrum Michx. Rough Hawkweed
Common. Sterile dry soil, old fields, open woods. Fl. 8(2).

H. venosum L. var. **nudicaule** (Michx.)Farw.
Rattlesnake-weed
Occasional. Dry rich soil, dry south-facing slopes in open woods. Fl. 6(1-3). Occasional in southern half of county, uncommon in northern half.

HYPOCHAERIS

***H. radicata** L. Cat's-ear
Uncommon. Fields. Fl. 6(3)-9. Sandisfield: Walters 1914 (NEBC). Introduced from Europe.

INULA

***I. helenium** L. Elecampane
Occasional. Moist soil, fields, roadsides. Fl. 7(3-4). Introduced from Europe.

KRIGIA

K. virginica (L.) Willd. Dwarf Dandelion
Frequent. Dry open sandy soil. Fl. 5(4)-6(4). In southernmost towns only.

LACTUCA

L. biennis (Moench) Fern. Wild Blue Lettuce
Frequent, moist soil on banks, borders of woods. Fl. 7(4)-8(3).

L. canadensis L. var. **canadensis** L. Yellow Wild Lettuce
(Including var. **integrifolia** (Bigel.) Torr. & Gray)
Common. Roadsides, old fields. Fl. 7(3)-8(2).

L. canadensis L. var. **latifolia** Kuntze Yellow Wild Lettuce
Common. Roadsides, dry fields. Fl. 7.

L. hirsuta Muhl. var. **sanguinea** (Bigelow) Fern.
Hairy Wild Lettuce
Uncommon. Dry open woods. Fl. 7-8.

***L. muralis** (L.) Gaertn. Wall Lettuce

Uncommon. Moist shaded woods along road. Fl. 7(2-3). Lenox, Pittsfield, Sheffield. Introduced from Europe. First discovered in the county by the author in 1990.

***L. serriola** L. Prickly Lettuce
(**L. scariola** L.)
Common. Waste areas. Fl. 8. Introduced from Europe.

LAPSANA

***L. communis** L. Nipplewort
Frequent. Moist soil, roadsides and borders of woods. Fl. 6(3)-7(4). Introduced from Europe.

LEONTODON

***L. autumnalis** L. Fall Dandelion
Common. Waste areas, roadsides, fields. 7(3)-8. Introduced from Europe.

LEUCANTHEMUM

***L. vulgare** Lam. Ox-eye Daisy
(**Chrysanthemum leucanthemum** L.)
Common. Fields, on roadsides, waste areas. Fl. 6(2)-8. Introduced from Europe.

MATRICARIA

***M. chamomilla** L. Wild Chamomile
Uncommon. Roadsides and pastures. Fl. 7(3)-8. Introduced from Europe.

***M. matricarioides** (Less.) Porter Pineapple-weed
(**M. discoidea** DC.)
Common. Bare gravel, waste areas, along roads. Fl. 6(3)-9. Introduced from Pacific states.

MEGALODONTA

M. beckii (Torr. ex Spreng.) Greene Water-marigold
Uncommon. In slow alkaline streams and ponds. Fl. 8(3). New Marlboro, Pittsfield, Sandisfield, West Stockbridge. Plate 7F.

ONOPORDUM

***O. acanthium** L. Scotch Thistle, Cotton Thistle
Uncommon. Hillside pasture. Fl. 7(3). Lanesboro: Winslow 1912 (NEBC). Introduced from Europe.

PETASITES

P. frigidus (L.) Fries var. **palmatus** (Ait.) Cronq. **T**
Sweet Coltsfoot
Rare. Forested fens, calcareous seeps. Fl. 4(2). Rarely flowers. Pittsfield, Sheffield, Williamstown.

PICRIS

***P. hieracioides** L. Ox-tongue
Uncommon. Dry fields. Fl. 6(2). Lenox: Hoffmann 1912 (NEBC). Introduced from Europe.

PRENANTHES

P. alba L. White Rattlesnake-root
Occasional. Moist rich woods and shaded calcareous ledges. Fl. 8(4)-9(1). West Stockbridge, Stockbridge.

P. altissima L. Rattlesnake-root
Common. Open woods, borders of woods. Fl. 8.

P. serpentaria Pursh **H** Lion's Foot, Gall-of-the-earth
Rare and historical. Dry thickets. Fl. 8-9.

P. trifoliolata (Cass.) Fern. Gall-of-the-earth
Frequent. Open dry woods. Fl. 7(4)-8.

RUDBECKIA

R. hirta L. var. **hirta** Black-eyed Susan
Uncommon. Lanesboro: Churchill 1916 (NEBC). This is not a good specimen of the native variety, which is now probably extirpated. Neither variety was listed by Dewey in 1829.

***R. hirta** L. var. **pulcherrima** Farw. Black-eyed Susan
Common. Fields, roadsides, waste areas. Fl. 6(4)-9. Introduced from farther west.

R. laciniata L. Tall Coneflower
Frequent. Wet soil, along rivers and streams, wet thickets. Fl. 7(3)-8.

SENECIO

S. aureus L. Golden Ragwort
Common. Wet soil, in marshes, swampy woods. Fl. 5(4)-6(2).

S. obovatus Muhl. ex Willd. Round-leaved Ragwort
Common. Open rocky banks, calcareous boulders and ledges. Fl. 5(2-4).

S. pauperculus Michx. Balsam Ragwort
Uncommon. Exposed calcareous ledges. Fl. 6(2). Lanesboro. First discovered in the county by B. A. Sorrie in 1984.

***S. vulgaris** L. Common Groundsel
Common. Waste areas. Fl. 5(4)-8. Introduced from Europe.

SILPHIUM

***S. perfoliatum** L. Cup-plant

Occasional. Moist or wet rich fields, roadsides. Fl. 8(2-4). Pittsfield, Otis. Introduced from farther west.

SOLIDAGO

S. arguta Ait. Toothed Goldenrod
Frequent. Dry open somewhat rich woods. Fl. 8(1-4).

S. bicolor L. White Goldenrod, Silver-rod
Common. Sterile soil, dry fields. Fl. 8(2)-9(2).

S. caesia L. Wreath Goldenrod, Blue-stem Goldenrod
Frequent. Dry or moist, somewhat rich open woods and borders of woods. Fl. 9(1-4). Hybridizes with **S. rugosa**.

S. canadensis L. var. **canadensis** Canada Goldenrod
(Including var. **hargeri** Fern.)
Common. Dry fields, borders of woods. Fl. 8(3)-9(4).

S. canadensis L. var. **scabra** (Muhl.) Torr. & Gray
(**S. altissima** L.) Tall Goldenrod
Common. Fields, borders of woods. Fl. 9.

S. flexicaulis L. Zig-zag Goldenrod
Frequent. Moist rich woods, borders of rich woods. Fl. 9.

S. gigantea Ait. Late Goldenrod
(Including **S. gigantea** Ait. var. **serotina** (Ait.) Cronq.)
Frequent. Moist soil, wet meadows, borders of brooks, fields. Fl. 8(3)-9(2).

S. glutinosa Nutt. ssp. **randii** (Porter) Cronq. **E**
(**S. simplex** Kunth ssp. **randii** (Porter) Ringius)
Rand's Goldenrod
Rare. On open ridges and ledges. Fl. 8(3)-9. A northern species. Sheffield, Mt. Washington.

S. hispida Muhl. ex Willd. Hispid Goldenrod
(**S. bicolor** L. var. **concolor** Torr. & Gray)
Uncommon. Open rocky woods. Fl. 8(3-4). Stockbridge: Hoffmann 1904 (NEBC). Hoffmann said it was frequent in the southern part of the valley, but it seems to be absent at present.

S. juncea Ait. Early Goldenrod
Common. Fields, roadsides. Fl. 7(4)-8(4).

S. macrophylla Pursh **T** Large-leaved Goldenrod
Rare. In open woods and along roads at elevations over 792 m. (2600′). Fl. 7(4)-8(4). Adams, Florida, New Ashford, North Adams, Williamstown. A northern species, common on summit of Mt. Greylock.

S. nemoralis Ait. Gray Goldenrod
Common. Sterile soil, dry open fields and ridges. Fl. 8(3)-9.

S. odora Ait. Sweet Goldenrod
Uncommon. Dry rocky woods and thickets. Fl. 8(1). Sheffield: Hoffmann 1912 (NEBC).

S. patula Muhl. ex Willd. Square-stem Goldenrod
Frequent. Somewhat to quite calcareous swamps. Fl. 9.

S. puberula Nutt. Downy Goldenrod
Frequent. Open dry ridges and rocky summits. Fl. 8(4)-9.

S. purshii Porter Bog Goldenrod
(**S. uliginosa** Nutt. var. **linoides** (Torr. & Gray) Fern.)
Occasional. Calcareous wet meadows and fens. Fl. 8(4)-9. Lenox, Monterey, New Marlboro, Richmond, West Stockbridge, Williamstown, Sheffield.

S. rigida L. **H** Stiff Goldenrod
Historical. In dry open woods, limestone ledges. Fl. 8. Sheffield: Churchill 1920 (NEBC). Hoffmann noted one sizable station, which unfortunately has not been rediscovered.

S. rugosa Mill. var. **rugosa** Rough Goldenrod
Common. Dry old fields, roadsides, wet woods. Fl. 8(1)-9(1).

S. rugosa Mill. var. **villosa** (Pursh) Fern.
Uncommon. Edges of woods, open woods. Mt. Greylock. Fl. 8(4)-9(1). Adams, New Ashford.

***S. sempervirens** L. Seaside Goldenrod
Uncommon. Salty soil along Massachusetts Turnpike. West Stockbridge: Mehrhoff 13980 1990 (CONN). First discovered in the county by L. Mehrhoff in 1990.

S. speciosa Nutt. Showy Goldenrod
Occasional. Calcareous open woods, borders of woods. Fl. 9. Sheffield. Alford.

S. squarrosa Muhl. Stout Goldenrod
Occasional. Dry open soil along streams and rivers, in dry clearings. Fl. 9(1-2). Deerfield R., Florida; Great Barrington, Sheffield. Bash Bish Brook, Mt. Washington.

S. uliginosa Nutt. Swamp Goldenrod
Occasional. Open wet meadows. Fl. 8(4)-9.

S. ulmifolia Muhl. ex Willd. Elm-leaf Goldenrod
Uncommon. Dry calcareous open woods. Fl. 8(4). New Marlboro. Sheffield.

SONCHUS

***S. arvensis** L. ssp. **arvensis** Sow-thistle
Common. Waste areas, roadsides. Fl. 7(4)-8. Introduced from Europe.

***S. arvensis** L. ssp. **uliginosus** (Bieb.) Nyman Sow-thistle
Occasional. Moist open soil. Fl. 9(1). Introduced from Europe.

***S. asper** (L.) Hill Spiny Sow-thistle
Common. Waste areas, abandoned fields. Fl. 8. Introduced from Europe.

***S. oleraceus** L. Annual Sow-thistle
Frequent, in waste areas. Fl. 7(4)-8. Introduced from Europe.

TANACETUM

***T. vulgare** L. Tansy
Common. Waste areas, roadsides, gravelly stream banks. Fl. 7(3)-8. Introduced from Europe.

TARAXACUM

***T. laevigatum** (Willd.) DC. Red-seeded Dandelion
Occasional. Fields, open soil. Fl. 5(4)-6(1). Hancock: <u>Bean & White</u> 1919 (NEBC). Introduced from Europe.

***T. officinale** Weber Brown-seeded Dandelion
Common. Lawns, fields, roadsides, waste areas. Fl. 4(3)-9. Introduced from Europe.

TRAGOPOGON

***T. dubius** Scop. Thickstem Goat's-beard
Uncommon. Waste areas. Fl. 8. Pittsfield, Sheffield. Introduced from Europe. First discovered in the county by H. E. Ahles in 1973.

***T. pratensis** L. Yellow Goat's-beard
Common. Fields, roadsides. Fl. 6(2) and sporadically through summer. Introduced from Europe.

TUSSILAGO

***T. farfara** L. Coltsfoot
Common. Moist roadsides, gravelly river banks. Fl. 3(4)-4. Introduced from Europe.

XANTHIUM

X. strumarium L. var. **glabratum** (DC.) Cronq. Sandbar Clotbur
Occasional. Alluvial soil. Fl. 9. Great Barrington, Stockbridge, Sheffield.

X. strumarium L. var. **canadense** (P. Mill.) Torr. & Gray Clotbur
Frequent. Silt and gravel bars in rivers, open alluvial soil. Fl. 9.

LILIOPSIDA (MONOCOTYLEDONS)

ALISMATACEAE

ALISMA

A. triviale Pursh — Large Water-plantain
(**A. plantago-aquatica** L. var. **americanum** Schultes & Schultes)
Occasional. In rich alluvial marshes. Fl. 8(1).

A. subcordatum Raf. — Lesser Water-plantain
(**A. plantago-aquatica** L. var. **parviflorum** (Pursh) Torrey)
Frequent. In open marshes. Fl. 7(3)-8(2).

SAGITTARIA

S. cuneata Sheldon **E** — Wapato, Arrowhead
Rare. In oxbow and backwater ponds along the Housatonic River. Fl. 6(3-4). Fr. 8(2)-9(3). Great Barrington, Lenox. Pittsfield, Sheffield.

S. graminea Michx. — Grass-leaf Sagittaria
Frequent. In wet peaty soil at edges of marshes, or in shallow water in mud or gravel bottom. Fl. 7(3)-10(1).

S. latifolia Willd. var. **latifolia** — Arrowhead, Duck Potato
Common. In marshes, wet meadows, pond shores. Fl. 7(4)-8(2). Fr. 8(-4).

HYDROCHARITACEAE

ELODEA

E. canadensis Michx. — Waterweed, Elodea
Common. In shallow alkaline ponds and slow streams. Fl. 7(4)-8(1).

E. nuttallii (Planch.) St. John — Narrow Waterweed
Occasional. More acidic water than preceding species. Great Barrington, Monterey, Pittsfield. Otis. First discovered in the county by C. B. Hellquist in 1972.

VALLISNERIA

V. americana Michx. — Tape-grass, Wild Celery
Frequent. Rooted in shallow water of lakes and ponds. Fl. 8(1-2).

SCHEUCHZERIACEAE

SCHEUCHZERIA

S. palustris L. **T** — Pod-grass
Rare. In open sphagnum mat of bog. Fl. 7(1). Sheffield.

POTAMOGETONACEAE

POTAMOGETON

P. alpinus Balbis **H** — Northern Pondweed
Historical. Richmond: Robbins 1864 (GH). As Hoffmann noted, the specimen label reads: "e rivulo prope viam ferream."

P. amplifolius Tuckerman — Large-leaved Pondweed
Common. In lakes, ponds and slow streams. Fr. 8(2).

P. amplifolius X **P. illinoensis** — Pondweed
Uncommon. In pond. Pittsfield: Hellquist 14685 1980 (NASC).

P. bicupulatus Fern. — Hairlike Pondweed
(**P. capillaceus** Poir.)
Uncommon. In acidic ponds. Fr. 7(4). Mt. Washington. Savoy.

P. confervoides Reichenb. — Tuckerman's Pondweed
Uncommon. In acidic ponds. Fr. 7(1-3). Mt. Washington. Savoy.

*__P. crispus__ L. — Crisped Pondweed
Common. In alkaline ponds, lakes and rivers. Introduced from Europe.

P. epihydrus Raf. — Surface Pondweed
Occasional. In alkaline ponds and slow streams. Fr. 8(3).

P. foliosus Raf. — Leafy Pondweed
Common. In shallow ponds, river backwaters, slow streams. Fr. 7(1)-9(1).

P. friesii Rupr. **T** — Fries's Pondweed
Rare. In alkaline pond. Fr. 6(3-4). West Stockbridge.

P. gramineus L. — Grassleaf Pondweed
(Including var. **myriophyllus** Robbins)
Frequent. In lakes and ponds. Fr. 8(1).

P. hillii Morong **SC** — Hill's Pondweed
Rare. Locally frequent in cold small slow highly alkaline streams and occasional in ponds. Fr. 7(2)-8(1). Most central valley towns from Hancock south except Sheffield and New

Marlboro. Berkshire County and western Vermont have a majority of the stations world-wide (Hellquist, 1984).

P. illinoensis Morong Shining Pondweed
Frequent. In alkaline water of slow streams and ponds. Fr. 8(3).

P. illinoensis X **P. nodosus**
Uncommon. In somewhat calcareous pond. Great Barrington. A probable hybrid of these two species.

P. natans L. Floating Pondweed
Common. In marshes, ponds, lakes and rivers. Fr. 7(4).

P. nodosus Poir. Long-leaved Pondweed
Occasional. In slow-moving alkaline waters. Fr. 7(4)-8(2).

P. oakesianus Robbins Oakes's Pondweed
Occasional. In acidic waters of ponds. Fr. 7(4). Monterey, Mt. Washington, Savoy, Washington. Sandisfield.

P. obtusifolius Mert. & Koch Blunt Pondweed
Uncommon. In cool waters of alkaline ponds. Fr. 8(2). Stockbridge. West Stockbridge.

P. ogdenii Hellquist & Hilton Ogden's Pondweed
(**P. hillii** X **P. zosteriformis**)
Uncommon. In alkaline waters.

P. pectinatus L. Sago Pondweed
Frequent. In the most alkaline ponds and lakes. Fr. 8(3).

P. perfoliatus L. Clasping Pondweed
(Including **P. perfoliatus** var. **bupleuroides** (Fern.) Farw.)
Occasional, in slow alkaline streams and lakes. Fr. 8(2).

P. praelongus Wulfen Whitestem Pondweed
Frequent. In alkaline ponds and lakes. Fr. 6(4).

P. praelongus X **P. richardsonii** Pondweed
Uncommon. In alkaline lakes. Lenox-Lee: Hellquist 4879 1972 (NASC).

P. pusillus L. var. **pusillus** Tiny Pondweed
Common. In alkaline waters of ponds and lakes. Fr. 8(3).

P. pusillus L. var. tenuissimus Mert. & Koch Pondweed
Common. In acidic and alkaline waters of marshes, backwaters of rivers, ponds and lakes.

P. richardsonii (Benn.) Rydb. Richardson's Pondweed
Occasional. In alkaline ponds, lakes and slow rivers. Fr. 7(3). Lenox, Monterey, New Marlboro, Pittsfield, Richmond. West Stockbridge.

P. robbinsii Oakes Robbins's Pondweed
Frequent. In lakes and ponds. Seldom fruits.

P. X**spathuliformis** (Robbins) Morong Pondweed
(**P. gramineus** X **P. nodosus**)
Uncommon. In rivers and lakes. Lenox: Hellquist 4978 1972 (NASC).

P. spirillus Tuckerm. Spiral Pondweed
Frequent. In moderately alkaline to acidic ponds and lakes. Fr. 7(4)-8(4).

P. vaseyi Robbins Vasey's Pondweed
Uncommon. Lake Garfield. Monterey: Hellquist 4448 1972 (NASC). First discovered in the county by C. B. Hellquist in 1972.

P. zosteriformis Fern. Flat-stem Pondweed
Common. In alkaline streams, rivers, ponds and lakes.

NAJADACEAE

NAJAS

N. flexilis (Willd.) Rostk. & Schmidt Common Naiad
Common. In ponds, lakes and streams, in shallow water. Fr. 8(2).

N. gracillima (A. Br.) Magnus Naiad
Uncommon. In acidic waters. Bog Pond, Savoy. Fr. 8(4). First discovered in the county by the author in 1994.

***N. guadalupensis** (Spreng.) Magnus Guadalupe Naiad
Uncommon. In an alkaline lake, Onota Lake. First discovered in the county by the author in 1992.

***N. minor** All. Lesser Naiad
Uncommon. In quiet waters of lakes. Fr. 8(1). Lanesboro, Lenox, Pittsfield, Stockbridge: Hellquist (9687) 1974 (NASC). Discovered for the first time in the state by C. B. Hellquist in 1974. Introduced from Europe.

ARACEAE

ACORUS

A. americanus (Raf.) Raf. Sweetflag
(**A. calamus** of American authors, not L.)
Frequent. In wet meadows and marshes. Fl. 6(3).

ARISAEMA

A. dracontium (L.) Schott **T** Green Dragon
Rare. In open moist alluvial woods. Fl. 6(3). Sheffield. Great Barrington: Sorrie et al. 2947 1985 (NEBC). Plate 8A.

A. triphyllum (L.) Schott ssp. **triphyllum** Jack-in-the-pulpit
(**A. atrorubens** (Ait.) Blume)
Common. In moist, usually rich woods and wet woods. Fl. 5(2).

A. triphyllum (L.) Schott ssp. **stewardsonii** (Britt.) Huttleston
Occasional. In cool woods. Fl. 5(4). Florida, Richmond, Savoy, West Stockbridge. Sheffield: Hoffmann 1920 (NEBC).

CALLA

C. palustris L. Wild Calla
Common. In bogs, margins of boggy ponds and marshes. Fl. 6(1-3).

ORONTIUM

O. aquaticum L. **T** Golden Club
Rare. In shallow water of sandy-bottomed pond. Fl. 5(3). This one population has been reduced to one plant in a vegetative state. Otis.

PELTANDRA

P. virginica L. Schott Arrow-arum
Common. In marshes, bogs, wet meadows, margins of rivers and streams. Fl. 7(1-3).

SYMPLOCARPUS

S. foetidus (L.) Salisb. Skunk Cabbage
Common, in wooded swamps and fens. Fl. 4(3). Scarce north of Lanesboro or on the Berkshire Plateau.

LEMNACEAE

LEMNA

L. minor L. Duckweed
Common. Floating on quiet water of ditches, marshes, ponds. Fl. 5(2)-9. Seldom seen flowering.

L. perpusilla Torr. Duckweed
Uncommon. In pond. Fl. 5-9. West Stockbridge: Evans, Fernald & Knowlton 1919 (NEBC).

L. trisulca L. Star Duckweed
Occasional. Floating just below surface in quiet waters of ponds, backwaters of rivers. Fl. 6-8. Lenox, New Marlboro, Sheffield, Stockbridge, West Stockbridge.

SPIRODELA

S. polyrhiza (L.) Schleid. Great Duckweed
Frequent. In quiet waters of ponds, rivers and marshes. Fl. 6-9.

WOLFFIA

W. columbiana Karst. Watermeal
Common. In quiet waters of ponds, marshes. Fl. 7-9. First discovered in the county by E. Hellquist in 1985.

W. borealis (Engelm.) Landolt Watermeal
(**W. punctata** Griseb.)
Uncommon. In quiet waters of ponds. Cheshire: E. Hellquist 368 1985 (NASC). First discovered in the county by E. Hellquist.

XYRIDACEAE

XYRIS

X. difformis Chapm. var. **difformis** Yellow-eyed Grass
Uncommon. In boggy pond. Fl. 7(4). Sheffield.

X. montana Ries Northern Yellow-eyed Grass
Uncommon. In open sphagnum bog mat, rarely on peaty margin of pond. Fl. 7(2-4). Mt. Washington, Savoy. Otis.

COMMELINACEAE

COMMELINA

*__C. communis__ L. var. **ludens** (Miq.) Pennell
Asiatic Dayflower
Frequent. In waste areas, gardens, roadsides.

ERIOCAULACEAE

ERIOCAULON

E. aquaticum (Hill) Druce Pipewort
(**E. septangulare** With.)
Occasional, in shallow water of acidic ponds, boggy ponds. Fl. 8(1).

JUNCACEAE

JUNCUS

J. acuminatus Michx. Sharp-fruited Rush
Uncommon. In wet soil and ditches. Fr. 7(2). New Marlboro, Lenox.

J. articulatus L. Jointed Rush
Common. On margins of small streams in water, marshes, wet meadows. Fr. 7(1)-8.

***J. balticus** Willd. var. **littoralis** Engelm. Brackish Rush
Uncommon. In calcareous marsh. Fr. 7(3). Cheshire, Lanesboro. First discovered in the county by the author in 1988. Probably introduced from east coast.

J. brachycephalus (Engelm.) Buch. Small-headed Rush
Frequent. In calcareous marshes. Fr. 8(1-4).

J. brevicaudatus (Engelm.) Fern. Narrow Rush
Common. In acidic boggy pond margins, bogs, open acidic soil. Fr. 8(3)-9(1).

J. bufonius L. var. **bufonius** Toad-rush
Common. In open disturbed soil, roadsides, wet mud. Fr. 6(3)-7.

J. canadensis Gay ex Laharpe var. **canadensis** Marsh Rush
Frequent. In acidic marshes, beaver ponds. Fr. 9.

J. dudleyi Wieg. Dudley's Rush
Frequent. In calcareous wet meadows. 7(4)-9(1).

J. effusus L. Soft Rush
(Including var. **solutus** Fern. & Wieg., var. **pylaei** (Laharpe) Fern. & Wieg., var. **decipiens** Buch., var. **compactus** Hoppe.)
Common. In marshes, wet meadows, bogs, borders of ponds. Fr. 8-9.

J. filiformis L. **T** Thread-rush
Rare. On boggy margins of ponds, swampy woods at higher elevations, over 1900′ (578 m), on Berkshire Plateau. Fr. 7. Florida, Peru, Savoy, Windsor.

***J. gerardii** Loisel. Black-grass
Uncommon. In disturbed soil of boat launch site, gas line, roadside. Fr. 8(1). Lee, Richmond, Stockbridge. First discovered in the county in 1984 by B. A. Sorrie. Native in coastal areas.

J. greenei Oakes & Tuckerm. Greene's Rush
Uncommon. In dry, sandy, sterile, open soil. Fr. 8(1-2). Pittsfield. First discovered in the county in 1988 by the author.

***J. inflexus** L. Meadow Rush
Uncommon. In wet meadow near gas pipeline. Fr. 8(3). Stockbridge. First discovered in New England in 1990 by the author. Introduced from Eurasia.

J. marginatus Rostk. var. **marginatus** Grass-rush
Occasional. In wet meadows. Fr. 7(3).

J. militaris Bigel. Bayonet-rush
Uncommon. In shallow water, at margin of sandy bottom pond. Fr. 8. Big Pond, Otis.

J. nodosus L. Knot-rush
Frequent. In calcareous wet meadows. Fr. 8(3).

J. pelocarpus Mey Pondshore Rush
Occasional. On peaty margins of bog ponds. Fr. 8.

J. secundus Beauv. Secund Rush
Uncommon. In dry, open, sandy soil on sandplain. Fr. 7(2-4). Alford. Sheffield.

J. tenuis Willd. var. **tenuis** Path rush
Common. In moist pastures, pathways, roadsides, sandy soil, flourishing in well-trodden soil. Fr. 7(3)-9(1).

***J. torreyi** Cov. Torrey's Rush
Uncommon. In disturbed moist soil. Fr. 7(4). Richmond, Williamstown. Native farther west. First discovered in the county in 1985 by the author.

LUZULA

L. acuminata Raf. Drooping Woodrush
Frequent. In rich woods, open woods and clearings. Fl. 5(1).

L. multiflora (Retz.) Lej. ssp. **multiflora** Common Woodrush
(**L. campestris** (L.) DC. var. **multiflora** (Retz.) Celak.)
Frequent. In open woods. Fl. 5(3).

L. parviflora (Ehrh.) Desv. ssp. **melanocarpa** (Michx.) Hamet-Ahti **E** Black-fruited Woodrush
Rare. In moist spruce-fir forest above 2800′ (854 m) on Mt. Greylock, often in the openings along trails. Fr. 7(1). Adams.

CYPERACEAE

BULBOSTYLIS

B. capillaris (L.) Clarke Sand-rush
(**Stenophyllus capillaris** (L.) Britt.)
Frequent. In sandy soil, along railroads, on sandplain. Fr. 8(2)-9(2).

CAREX

C. aenea Fern. Fernald's Hay Sedge
Occasional. In dry open woods, rocky woods, open rocky ridgetops. Fr. 6(2)-7(4). New Marlboro. Becket. Stockbridge/West Stockbridge.

C. aestivalis Curtis Summer Sedge
Occasional. In open woods on mountain ridges. Fr. 6(3).

C. albursina Sheldon White Bear Sedge
Frequent. In moist rich woods. Fr. 7(2).

C. alopecoidea Tuckerm. **T** Foxtail Sedge
Rare. In alluvial swales, floodplain forests. Cheshire, Lenox, North Adams, Sheffield, Williamstown. Great Barrington.

C. amphibola Steud. var. **rigida** (Bailey) Fern. Stiff Sedge
Uncommon. In rich woods. Sheffield. West Stockbridge: Sorrie 2056 1983 (NEBC).

C. amphibola Steud. var. **turgida** Fern. Gray Sedge
(**C. grisea** Wahl.)
Frequent. In rich alluvial soil and rich woods. Fr. 5(4).

C. annectens (Bickn.) Bickn. Yellow Fox Sedge
Occasional. In open moist meadows. Fr. 7(4). Becket, Hinsdale, Sheffield, W. Stockbridge, Williamstown. Hoffmann listed this species as frequent, but it is seldom found.

C. appalachica Webber and Ball Appalachian Sedge
(**C. radiata** Mackenzie; **C. rosea** var. **radiata** Boott)
Frequent. On rich rocky wooded slopes. Fr. 6(3-4). (See Webber and Ball, 1984).

C. aquatilis Wahl. var. **aquatilis** Water Sedge
Occasional. In open calcareous marshes, a major component of lake basin graminoid fens. Fr. 7(1).

C. arctata Boott Woodland Drooping Sedge
Frequent, in moist woods, clearings. Fr. 6(3).

C. argyrantha Tuckerm. Silvery Sedge
Occasional. In dry woods, in rocks on open ridges. Fr. 6(4)-8(1). Great Barrington. Mt. Washington.

C. artitecta Mackenzie Sedge
(**C. albicans** Willd. var. **albicans**)
Uncommon. On calcareous outcrops. Fr. 5(4). Sheffield. Stockbridge: Hoffmann 1920 (NEBC).

C. atlantica Bailey ssp. **atlantica** Atlantic Prickly Sedge
(**C. atlantica** Bailey var. **incomperta** F. J. Herm.)
Uncommon. Open Sphagnum mat of bogs, margins of acidic ponds. Fr. 6(2). Becket, New Marlboro. Sandisfield. Mt. Washington.

C. atlantica Bailey ssp. **capillacea** (Bailey) Reznicek
(**C. howei** Mackenzie) Threadstem Prickly Sedge
Occasional. On peaty margins of sphagnum bogs, boggy marshes. Fr. 6(3). On Berkshire Plateau and Taconic Range. For treatment of Stellulatae, see Reznicek & Ball (1980).

C. aurea Nutt. Golden Sedge
Frequent, in calcareous wet meadows and seeps. Fr. 7(2).

C. baileyi Britt. **E** Bailey's Sedge
Rare. In moist calcareous soil, wooded seepage area, edge of swamp. Fr. 7(3). New Marlboro, Cheshire.

C. bebbii (Bailey) Fern. Bebb's Broom Sedge
Common. In marshes, wet meadows, open woods, margins of streams. Fr. 7(1-3).

C. blanda Dewey Woodland Sedge
(**C. laxiflora** Lam. var. **blanda** (Dewey) Boott)
Frequent. In rich woods. Fr. 5(4). Most towns in the valley.

C. brevior (Dewey) Mackenzie ex Lunell Short Fescue Sedge
(**C. festucacea** Willd. var. **brevior** (Dewey) Fern.)
Occasional. In dry sandy open soil, on sandplain, sand pits. Fr. 7(3-4). Great Barrington. Sheffield.

C. bromoides Willd. Brome Sedge
Common. In open marshes, wooded swamps, often in somewhat calcareous soil. Fr. 6(4).

C. brunnescens (Pers.) Poir. ssp. **sphaerostachya** (Tuckerm.) Kalela Sedge
Occasional. In open woods and wet woods on mountain summits and on Berkshire Plateau. Fr. 7(1). Common on Mt. Greylock.

C. bushii Mackenzie **E** Bush's Sedge
Rare. In moist calcareous wet meadow or open dry sandy soil. Fr. 6(4)-8. Lenox, Sheffield, Williamstown. First discovered in the county by the author in 1986.

C. buxbaumii Wahl. Buxbaum's Sedge
Uncommon. In calcareous marsh. Fr. 6(3). Egremont: Hoffmann 1920 (NEBC).

C. canescens L. ssp. **canescens** Silvery Bog Sedge
Frequent. In bogs and marshes. Fr. 6(4)-7(2).

C. canescens L. ssp. **disjuncta** (Fern.) Toivonen
Frequent. In bogs and marshes. Fr. 6(4)-7(2).

C. castanea Wahlenb. **E** Chestnut-colored Sedge
Rare. In calcareous wet meadow, margin of boggy pond. Fr. 5(3)-6(1). Pittsfield, Lenox. First discovered in Massachusetts by C. Caljouw and T. Rawinski in 1984.

C. cephaloidea (Dewey) Dewey Thin-leaved Sedge
(**C. sparganioides** Willd. var. **cephaloidea** (Dewey) Carey)
Occasional, in alluvial meadows and woods. Fr. 6(1).

C. cephalophora Muhl. ex Willd. Oval-headed Sedge
Occasional. In dry, rich woods. Fr. 7(4).

C. chordorrhiza Ehrh. ex L.f. **E** Creeping Sedge
Rare. In open graminoid fens. Fr. 6(4). Richmond. First discovered in Massachusetts by C. Caljouw and T. Rawinski in 1985.

C. communis Bailey Colonial Sedge
Common. In deciduous woods, sometimes rich woods, occasional in dry sandy soil. Fr. 6(1-4).

C. comosa Boott Bristly Sedge
Common. In open marshes. Fr. 7(2).

C. conoidea Schk. Field Sedge
Frequent, in calcareous wet meadows. Fr. 6(4)-7(2).

C. crawfordii Fern. Crawford's Sedge
Occasional. On open rocky mountain ridges, sterile fields. Fr. 7(4). Adams, Clarksburg, Florida, Lanesboro, Savoy. More common in northern Berkshire Plateau.

C. crinita Lam. Awned Sedge
Occasional, in rich alluvial soil. Fr. 8-9. New Marlboro. Sheffield.

C. cristatella Britt. Crested Sedge
Frequent. In marshes along rivers and streams. Fr. 6(4)-7(3).

C. cryptolepis Mackenzie Small Yellow Sedge
Occasional. In somewhat disturbed calcareous meadows and marshes, limy rubble. Fr. 6(4)-7(4).

C. cumulata (Bailey) Fern. Crowded Sedge
Occasional. In clearings, bogs, dry sandy soil. Fr. 7(2).

C. davisii Schwein. & Torr. **E** Davis's Sedge
Rare. In alluvial soil along Housatonic R. Fr. 6(1-3). Sheffield. Hoffmann (1922) stated the type locality was in Sheffield, location unknown.

C. debilis Michx. var. **interjecta** Bailey
Southern Stalked Sedge
Uncommon. In open woods, open mountain summits. Fr. 6(3). Lanesboro: Churchill 1916 (NEBC).

C. debilis Michx. var. **rudgei** Bailey Northern Stalked Sedge
Common. In deciduous woods, rich woods, meadows. Fr. 6(3)-7(4).

C. deflexa Hornem. Mountain Sedge
Uncommon. Rocky shores of Deerfield R. Fr. 6(1). Florida: Hoffmann 1920.

C. deweyana Schwein. Dewey's Sedge
Common. In dry open and rich rocky woods. Fr. 6(3)-7(1).

C. diandra Schrank Panicled Sedge
Frequent. In open, graminoid fens. Fr. 6(3)-7(1).

C. digitalis Willd. Slender-Spiked Woodland Sedge
Frequent. In deciduous woods. Fr. 6(4)-7(2).

C. disperma Dewey Two-seeded Bog Sedge
Frequent. In swampy woods, wet meadows. Fr. 6(2).

C. eburnea Boott Thread-leaved Sedge
Occasional. In dry, open rocky calcareous outcrops or open woods, often occurring in large patches. Fr. 6(1-3).

C. echinata Murr. var. **echinata** Prickly Sedge
(**C. stellulata** Gooden.)
Occasional. In peaty soil on bog margins, margins of acidic marshes. Fr. 6(2)-7(3). Primarily on the Berkshire Plateau.

C emmonsii Dewey Emmons' Sedge
(**C. albicans** Willd. var. **emmonsii** (Dewey) J. Rettig
Uncommon. In dry open rocky woods. Fr. 5(3). Sheffield, Stockbridge, West Stockbridge, Mt. Washington.

C. flava L. var. **flava** Yellow Sedge
Common. In calcareous wet meadows, seeps, marshes. Fr. 6(2-4). A good indicator of at least some calcareous influence.

C. folliculata L. Long-culmed Sedge
Frequent. On shores of acidic ponds, beaver marshes, wet acidic woods. Fr. 6(3)-8(1). Mainly on the Berkshire Plateau.

C. formosa Dewey **T** Handsome Sedge
Rare. In rich moist woods, wooded seepage slopes, often at edge of calcareous marsh. Fr. 6. Egremont. Stockbridge.Type locality is in Stockbridge, location unknown (Hoffmann, 1922).

C. gracillima Schwein. Slender Sedge
Common. In moist woods, meadows, clearings. Fr. 6(1)-7.

C. granularis Muhl. ex Willd. var. **granularis** Meadow Sedge
Uncommon. In wet meadows. Fr. 6(2-3). Lanesboro: Batchelder 1916 (NEBC).

C. granularis Muhl. ex Willd. var. **haleana** (Olney) Porter
Hale's Meadow Sedge
Frequent. In calcareous wet meadows, seepy pastures, moist woods. Fr. 6(2-3).

C. grayi Carey **T** Gray's Sedge
Rare. In rich alluvial soil in floodplain forests along Housatonic R. Fr. 7. Lenox. Sheffield.

C. gynandra Schwein. Northern Awned Sedge
(**C crinita** Lam. var. **gynandra** (Schwein.) Schwein. & Torr.)
Common. In acidic or circumneutral marshes, along streams, in swampy spruce-fir forests, wet woods. Fr. 6(4)-7(4).

C. X **hartii** Dewey Sedge
(**C. lurida** X **C. retrorsa**)
Uncommon. In alluvial soil. Sheffield: Churchill 1920 (NEBC).

C. haydenii Dewey — Hayden's Sedge
Uncommon. In open marsh along brook and wet meadows Fr. 7(1). Savoy, Windsor.

C. hirsutella Mackenzie — Hirsute Sedge
(**C. complanata** Torr. & Hook. var. **hirsuta** (Bailey) Gleason)
Frequent. In dry or moist soil in pastures, clearings. Fr. 6(4)-7(3).

***C. hirta** L. — Hairy Sedge
Uncommon. In moist fields near old farm site. Fr. 7(1-2). Williamstown. Introduced from Europe. First found in the county by the author in 1985.

C. hirtifolia Mackenzie — Pubescent Sedge
Occasional. In moist rich woods and rich alluvial soil. Fr. 6(3-4).

C. hitchcockiana Dewey **SC** — Hitchcock's Sedge
Rare. In rich woods, near calcareous outcrops. Fr. 6(3-4). Lanesboro, Pittsfield, Richmond, Sheffield, West Stockbridge, Williamstown.

C. hystericina Muhl. ex Willd. — Porcupine Sedge
Common. In wet meadows, marshes, usually sites with a calcareous influence. Fr. 6(1-3).

C. interior Bailey — Inland Prickly Sedge
(**C. scirpoides** Schkuhr.)
Frequent. Wet calcareous soil, fens, marshes, wet meadows. Fr. 6(1-3).

C. intumescens Rudge — Bladder Sedge
Frequent. In moist woods, swampy woods. Fr. 6(4)-10(1).

C. lacustris Willd. — Lakeside Sedge
Common. In marshes, river marshes, open fens. Fr. 6(2-4).

C. laevivaginata (Kukenth.) Mackenzie — Smooth-sheathed Sedge
Occasional, in moist to wet rich woods, open fens. Fr. 6(3)-7.

C. lanuginosa Michx. — Broad Woolly-fruited Sedge
Occasional, in calcareous wet meadows, marshes and fens. Fr. 6(2)-7(1).

C. lasiocarpa Ehrh. var. **americana** Fern. — Slender Woolly-fruited Sedge
Frequent. In bogs and fens, boggy margins of slow streams and ponds. Fr. 6(2)-7(2).

C. laxiculmis Schwein. — Spreading Woodland Sedge
Frequent. In moist rich woods. Fr. 6(1-4).

C. laxiflora Lam. — Loose-flowered Woodland Sedge
Common. In moist rich woods. Fr. 6(1-2).

C. lenticularis Michx. **T** — Shore Sedge
Rare. On gravelly or rocky shores of Deerfield River and acidic ponds on Berkshire Plateau, usually in water. Fr. 7(3-4). Otis, Savoy, Florida.

***C. leporina** L — Hare's-foot Sedge
Uncommon. In pastures. New Marlboro. Introduced from Europe.

C. leptalea Wahl. — Delicate Sedge
Frequent. In calcareous seeps, fens, marshes, wet meadows, bogs and wet woods. Fr. 7.

C. leptonervia Fern. — One-nerved Woodland Sedge
Frequent. In moist woods. Fr. 6(3-4).

C. limosa L. — Mud Sedge
Uncommon. On open sphagnum mat of bogs and sphagnum tussocks in fens. Fr. 6(4)-7(3). New Marlboro, Sheffield, Stockbridge, Washington, Otis.

C. lucorum Link — Long-beaked Pennsylvania Sedge
(**C. pensylvanica** Lam. var. **distans** Peck)
Occasional. In dry rocky woods, upper rocky slopes of mountains. Fr. 5(3)-6. Great Barrington, Mt. Washington, Sheffield. West Stockbridge: Sorrie & Holland 3050 1985 (NEBC).

C. lupulina Willd. — Hop Sedge
Frequent. In wet woods, marshes, swales along rivers and streams. Fr. 7(3)-8.

C. lurida Wahl. — Sallow Sedge
Common. Widespread in marshes, wet meadows, wet places in open woods. Fr. 6-8.

C. michauxiana Boekl. **E** — Michaux's Sedge
Rare. On peaty margin of pond. Fr. 6(3). Savoy. First discovered in Massachusetts by H. E. Ahles in 1972.

C. muhlenbergii Schk. — Muhlenberg's Sedge
Uncommon. In dry open calcareous rubbly soil. Fr. 7(2). Sheffield.

C. normalis Mackenzie — Big Straw Sedge
Occasional, in dry open woods, clearings, edge of marsh. Fr. 6(3-4).

C. novae-angliae Schwein. — New England Sedge
Occasional. In cool, moist woods, open ledges, at higher elevations. Fr. 6(2-3). Adams, Becket, North Adams, New Marlboro,

Savoy. Lanesboro. Type locality is on Mt. Greylock (Hoffmann, 1922).

C. oligocarpa Schk. **H** Rich Woods Sedge
Rare and historical. In rich woods. New Marlboro, Stockbridge.

C. ormostachya Wieg. Necklace Woodland Sedge
Occasional. In rich woods, banks, margins of fens. Fr. 6(3-4). North Adams, Sandisfield, Savoy, Stockbridge. West Stockbridge.

C. pallescens L. Shiny Green Sedge
Frequent. In moist meadows, old fields, pastures. Fr. 6(4).

*__C. panicea__ L. Millet Sedge
Uncommon, in seepy margin of small pond in pasture. Fr. 7(4). Williamstown. First discovered in the county by the author in 1985. Introduced from Europe.

C. pauciflora Lightf. **H** Few-flowered Sedge
Rare and historical. On sphagnum margins of peat bogs. Fr. 6(4). New Marlboro: <u>Hoffmann</u> 1912 (NEBC).

C. paupercula Michx. var. **irrigua** (Wahl.) Fern.
(**C. magellanica** Lam. ssp. **irrigua** (Wahl.) Hulten)
Pendant Bog Sedge
Occasional. In sphagnum in bogs near shrub zone. Fr. 6(4)-7(2). Becket, Clarksburg, Otis, Peru.

C. paupercula Michx. var. **pallens** Fern. Pendant Bog Sedge
(**C. magellanica** Lam. ssp. **magellanica**)
Occasional. In sphagnum at edge of bog mat, in bog forest. Fr. 6(4)-7(2). Stockbridge, Washington. New Marlboro.

C. pedunculata Willd. Long-stalked Sedge
Frequent. In dry rich woods. Fr. 5(1).

C. pensylvanica Lam. Pennsylvania Sedge
Common. Dry open woods, often forming large patches. Fr. 5(2).

C. plantaginea Lam. Plantain-leaf Sedge
Frequent. In moist rich woods, often abundant. Fr. 5(1).

C. platyphylla Carey Broad-leaf Sedge
Common. In dry or moist rich woods. Fr. 6(3)-8(2).

C. prairea Dewey Prairie Sedge
Occasional. In open graminoid fens, calcareous marshes. Fr. 6(2).

C. prasina Wahl. Twisted Drooping Sedge
Frequent. In wooded swamps, wet woods. Fr. 5(4)-7(4).

C. projecta Mackenzie Beaded Broom Sedge
Frequent. In dry or moist open woods, clearings, wooded swamp. Fr. 7(2)-8(1).

C. pseudocyperus L. Galingale Sedge
Occasional. In calcareous marshes and open fens. Fr. 6(2-4).

C. radiata (Wahl.) Small Star Sedge
(**C. rosea** Willd. var. **radiata** (Wahl.) Dewey; **C. rosea** var. **minor** Boott; **C. rosea** Mackenzie)
Occasional. On rich rocky wooded slopes. Fr. 6(3-4). Becket, Great Barrington, Hancock, Williamstown. New Marlboro. (See Webber and Ball, 1984).

C. retrorsa Schwein. Hooked Sedge
Common. In river marshes, swamps in woods, calcareous marshes. Fr. 6(4)-8(4).

C. rosea Schkuhr ex Willd. Big Star Sedge
(**C. convoluta** Mackenzie)
Frequent. In rich or rocky woods. Fr. 6(3). (See Webber and Ball, 1984).

C. rostrata Stokes var. **rostrata** Bottle Sedge
Uncommon. In open wet calcareous meadows, marshes. Fr. 7(1).

C. rostrata Stokes var. **utriculata** (Boott) Bailey
Beaked Sedge
Frequent. In marshes, wet meadows. Fr. 7(1).

C. rugosperma Mackenzie Sand Sedge
Occasional. In open rocky summits, dry sandy soil. Fr. 5(2)-6(4). Sheffield: <u>Bean</u> & <u>Fernald</u> 1919 (NEBC).

C. scabrata Schwein. Scabrous Sedge
Common. In wet woods, swampy woods along streams. Fr. 6(1)-7. Type locality in Berkshire County (Hoffmann, 1922).

C. schweinitzii Dewey ex Schwein. **E** Schweinitz's Sedge
Rare. In calcareous open fens. Fr. 6(4)-7(2). Cheshire, Lanesboro, West Stockbridge, Williamstown. Type locality in Williamstown (Hoffmann, 1922).

C. scoparia Schk. ex Willd. Broom Sedge
Common. In open marshes, dry disturbed areas, edges of woods and trails. Fr. 6(3)-7.

C. seorsa Howe Swamp Prickly Sedge
Uncommon. In sphagnum margins of bogs, open swamps. Fr. 5(3)-6. Becket, Sheffield: <u>Bean</u> & <u>Fernald</u> 1919 (NEBC).

C. sparganioides Willd. Bur Sedge
Frequent. In rich woods. Fr. 6(4)-7(2).

C. sprengelii Dewey ex Spreng. Long-beaked Sedge

Frequent. In alluvial soil, moist margins of woods. Fr. 5(3). Type locality in Sheffield (Hoffmann, 1922).

C. sterilis Willd. **T** Dioecious Sedge
Rare. In open graminoid fens. Fr. 6(4)-7(1). Egremont, Sheffield, West Stockbridge.

C. stipata Muhl. Awl-fruited Sedge
Common. In open graminoid marshes, wet meadows, river swales, swampy woods. Fr. 6(2)-7(2).

C. stricta Lam. Tussock Sedge
Common. In open marshes, borders of ponds and bogs, wet meadows. Fr. 6(3)-7(2). (See Standley 1989).

C. swanii (Fern.) Mackenzie Swan's Sedge
Common. In dry fields, dry disturbed soil. Fr. 7(4)-8(2).

C. tenera Dewey Slender Straw Sedge
(Including var. **echinodes** (Fern.) Wieg.)
Frequent. In alluvial soil along rivers, streams, borders of marshes. Fr. 6(2)-7(1).

C. tetanica Schk. **SC** Fen Sedge
Rare. In calcareous wet meadows and fens. Fr. 6(3)-8. Egremont, Lanesboro, Lee, Lenox, Richmond, Stockbridge, West Stockbridge, Sheffield.

C. tincta Fern. Brown Straw Sedge
Occasional. In dry fields, roadsides, in Deerfield River valley. Fr. 6(2)-7.

C. tonsa (Fern.) Bickn. Stiff-leaved Sand Sedge
Uncommon. In dry sandy soil in sandplain, or crevices in rocky summits. Fr. 5(1)-6(3). Sheffield, Great Barrington.

C. torta Boott Riverside Sedge
Common. Among rocks in and bordering fast-moving rivers and streams, sometimes inundated. Fr. 6(4)-7(3).

C. tribuloides Wahl. Blunt Broom Sedge
Frequent. In open marshes, wet meadows, beaver meadows, river swales. Fr. 6(2)-10(1).

C. trichocarpa Schk. **T** Hairy-fruited Sedge
Rare. In alluvial soil, in wet meadows and swales along the Hoosac River. Fr. 6(4)-7(2). Adams, Cheshire, North Adams. Williamstown.

C. trisperma Dewey var. **billingsii** Knight
Three-seeded Bog Sedge
Occasional. In peaty margins of peat bogs. Fr. 8. Becket, New Marlboro. Otis.

C. trisperma Dewey var. **trisperma** Three-seeded Bog Sedge
Frequent. In boggy woods, wet woods. Fr. 8.

C. tuckermanii Dewey **E** Tuckerman's Sedge
Rare. In rich alluvial soil along Konkapot and Housatonic Rivers, deep soil in ravine. Fr. 8-9(2). New Marlboro, Sheffield, Stockbridge.

C. typhina Michx. **H** Cat-tail Sedge
Rare and historical. In shaded swale near Housatonic River, alluvial soil. Fr. 7(1)-8. Sheffield: Hoffmann 1913 (NEBC).

C. vesicaria L. Inflated Sedge
(Including vars.**jejuna** Fern., **monile** (Tuckerman) Fern.)
Common. In open marshes, river swales. Fr. 7.

C. virescens Willd. Ribbed Sedge
Common. In dry, rocky woods. Fr. 6(2)-7(3).

C. viridula Michx. ssp. **viridula** Little Green Sedge
(**C. oederi** Retz ssp. **viridula** (Michx.) Hultén)
Occasional. In moist somewhat disturbed situations, old marble quarry, railroad tracks, ditches. Fr. 8(2)-9(3).

C. vulpinoidea Michx. Fox Sedge
Common. In wet meadows, moist soil on margins of ponds, streams, disturbed areas. Fr. 7-8(2).

CLADIUM

C. mariscoides (Muhl.) Torr. Twig-rush
Uncommon. At edge of sphagnum bog mats, boggy margins of ponds, graminiod basin fen. Fr. 8-10(3). Otis, Sheffield, Stockbridge, Sandisfield.

CYPERUS

C. aristatus Rottb. Awned Flatsedge
(**C. squarrosus** L.; **C. inflexus** Muhl.)
Occasional. On sandy to muddy upper banks of Housatonic River. Fr. 8(4)-9. Sheffield.

C. bipartitus Torrey Red-scaled Flatsedge
(**C. rivularis** Kunth)
Frequent. Sandy to muddy open shores of ponds, streams.

C. diandrus Torr. Strand Flatsedge
Occasional. On exposed sandy or muddy shores of ponds, rivers and marshes. Fr. 8(3)-9.

C. erythrorhizos Muhl. Redroot Flatsedge
Uncommon. On rocky shore of lake and river marsh along Housatonic R. Fr. 8(3)-9. Monterey, Sheffield. First discovered in the state by G. Tucker in 1985.

C. esculentus L. Yellow Nut-grass
Common. On river banks, wet meadows, pastures and as a garden weed. Fr. 8(3)-9.

C. filiculmis Vahl — Sand Flatsedge
Occasional. In dry open sandy soil. Fr. 7(3)-9(1).

C. houghtonii Torr. — Houghton's Flatsedge
Uncommon. On exposed rock ledges. Fr. 8(2)-9. Mt. Washington, Alford-West Stockbridge.

C. strigosus L. — Galingale
Common. In wet meadows and pastures, muddy banks of streams. Fr. 8(3)-9(2).

DULICHIUM

D. arundinaceum (L.) Britt. — Three-way Sedge
Common. In marshes, borders of ponds. Fr. 8.

ELEOCHARIS

E. acicularis (L.) Roemer & Schultes — Needle Spikesedge
Frequent. On muddy or peaty shores forming mats, often submersed and sterile. Fr. 7(4)-8.

E. elliptica Kunth — Orange-fruited Spikesedge
Frequent. On open wet soil in mud, clay, peat, and in disturbed areas. Fr. 7(2)-10(3).

E. erythropoda Steud. — Redfoot Spikesedge
Frequent. In calcareous wet meadows, marshes, also peaty margins of ponds. Fr. 7.

E. intermedia Schultes **T** — Mudflat Spikesedge
Rare. On exposed mud of marshes, backwaters, slow streams, abandoned beaver ponds where there is calcareous influence. Fr. 7-8. Most central valley towns.

E. obtusa (Willd.) Schultes — Softstem Spikesedge
Common. In moist or wet exposed soil, pond or stream margins, ditches. Fr. 7(2)-10(3).

E. olivacea Torr. var. **olivacea** — Olive Spikesedge
Occasional. On peaty margins and exposed mud of marshes and bogs. Fr. 8. Tyringham. Mt. Washington, Otis.

E. pauciflora (Lightf.) Link var. **fernaldii** Svenson **E**
(**E. quinqueflora** (F. X. Hartman) Schwarz)
Few-flowered Spikesedge
Rare. In exposed mud pools devoid of other vegetation in basin fen. Stockbridge. First discovered in the state by B. A. Sorrie and C. Caljouw in 1989.

E. smallii Britt. — Small's Spikesedge
Common. In open marshes along rivers, ponds. Fr. 7(4)-9.

ERIOPHORUM

E. alpinum L. — Alpine Cottongrass
(**Scirpus hudsonianus** (Michx.) Fern.)
Uncommon. In open peaty fens. Fr. 6(1)-7(2). Cheshire, Egremont, Lanesboro, Stockbridge, West Stockbridge.

E. gracile W. D. J. Koch **T** — Slender Cottongrass
Rare. In peat bogs, open graminoid fens. Fr. 6(2-3). Cheshire, Egremont, Monterey, Richmond, Sheffield, Stockbridge.

E. tenellum Nutt. — Rough Cottongrass
Occasional. On sphagnum mat of peat bogs. 7-8(2). Becket Otis, Sheffield, Washington.

E. vaginatum L. ssp. **spissum** (Fern.) Hultén — Hare's Tail
(**E. spissum** Fern.)
Uncommon. In sphagnum in openings in bog forest, peat bogs. Fr. 6(2-3). Becket, Savoy, Washington. New Marlboro. More uncommon than formerly, some habitats destroyed or flooded.

E. virginicum L. — Tawny Cottongrass
Frequent. In sphagnum mat of peat bogs. Fr. 8-9(2). Mostly on Berkshire Plateau.

E. viridicarinatum (Engelm.) Fern. — Fen Cottongrass
Common. In calcareous wet meadows, seeps. Fr. 6(1-4).

FIMBRISTYLIS

F. autumnalis (L.) Roemer & Schultes — Northern Fimbry
Uncommon. On sandy or gravelly shores of ponds and rivers. Fr. 8(2)-9. Richmond. Sheffield: Hoffmann 1912 (NEBC).

RHYNCHOSPORA

R. alba (L.) Vahl — White Beaksedge
Frequent. In sphagnum bogs, peaty margins of ponds. Fr. 8(3)-9(2).

R. capillacea Torr. **E** — Capillary Beaksedge
Rare. In open graminoid or shrub fens. Fr. 8. Pittsfield. Egremont: Sorrie 3586 1986 (NEBC). First discovered in the state by B. A. Sorrie in 1986.

R. capitellata (Michx.) Vahl — Brown Beaksedge
Frequent. In wet peaty or sandy shores, wet meadows and margins of marshes. Fr. 7(4)-8.

SCIRPUS

S. acutus Muhl. ex Bigel. — Hard-stem Bullsedge
Occasional. At edge of open water in fens, calcareous marshes. Fr. 7(4)-8.

S. atrocinctus Fern — Northern Bullsedge

Common. Wet meadows, marshes along ponds and rivers. Fr. 6(3)-7(3). Most common on Berkshire Plateau.

S. atrovirens Willd. Dark Green Bullsedge
Frequent. In wet meadows, borders of marshes. Fr. 7(4)-8.

S. cyperinus (L.) Kunth Wool-grass
Common. In marshes, wet meadows, pond shores. Fr. 8(1)-9.

S. expansus Fern. Spreading Bullsedge
Occasional. In calcareous marshes and shrub swamps. Fr. 8(4)-9. Stockbridge, New Marlboro. Monterey.

S. hattorianus Makino Meadow Bullsedge
Common. In wet meadows, moist soil in woods, along trails. Fr. 7(4)-8.

S. microcarpus Presl Barberpole Bullsedge
(**S. rubrotinctus** Fern.)
Common. In open marshes, wet meadows. Fr. 7(2)-8.

S. pedicellatus Fern. Stalked Bullsedge
(**S. cyperinus** (L.) Kunth var. **pedicellatus** (Fern.) Schuyler)
Occasional. In swamps along rivers. Fr. 7(4)-9. Clarksburg, Sheffield, Washington. Lenox.

S. pendulus Muhl. **SC** Pendulous Bullsedge
Rare. In calcareous wet meadows, seeps, fens. Fr. 7(2-4). Most towns in central valley region.

S. pungens Vahl Chairmaker's Rush, Three-square
(**S. americanus** Pers. has been misapplied to this taxon.)
Occasional. On sandy shores, margins of ponds and marshes. Fr. 8(3)-9(2). Otis, Stockbridge.

S. smithii Gray Smith's Clubsedge
Occasional. On open peaty margins of ponds and marshes. Fr. 8(2-4). Becket, Lenox, Mt. Washington,Tyringham.

S. subterminalis Torr. Water Clubsedge
Occasional. In shallow acidic boggy ponds. Fr. 8. Great Barrington, Monterey, Mt. Washington, Stockbridge, Otis.

S. tabernaemontanii Gmel. Soft-stem Bullsedge
(**S. validus** Vahl)
Frequent. In open marshes, margins of ponds and lakes. Fr. 7(4)-9(1).

S. torreyi Olney Torrey's Bullsedge
Uncommon. Marshy borders of ponds. Fr. 8(3)-9. Great Barrington, Clarksburg.

S. verecundus Fern. Bashful Bullsedge
(**S. planifolius** Muhl.)
Uncommon. In open dry rocky woods, steep south-facing slopes, somewhat rich soil. Fr. 5(2-4). Egremont, Mt. Washington, West Stockbridge.

POACEAE

AGROPYRON

***A. repens** (L.) Beauv. Quackgrass
Common. In fields, roadsides, lawns and gardens. Fr. 7-8. Introduced from Europe.

A. trachycaulum (Link) Malte ex Lewis var. **trachycaulum** (**Elymus trachycaulus** (Link) Gould ex Shinners ssp. **trachycaulus**) Slender Wheatgrass
Occasional. On calcareous outcrops, dry open wooded rocky ridgetops, calcareous marshes. Fr. 7(3).

A. trachycaulum (Link) Malte ex Lewis var. **glaucum** (Pease & Moore) Malte Slender Wheatgrass
(**Elymus trachycaulus** (Link) Malte ex Shinners ssp. **subsecundus** (Link) A. & D. Love)
Occasional. On calcareous outcrops, dry open wooded rocky ridgetops, calcareous marshes. Fr. 6(4)-7(3).

AGROSTIS

***A. capillaris** L. Rhode Island Bent
(**A. tenuis** Sibth.; **A. sylvatica** Huds.)
Common. In dry sandy soil on roadsides, fields, gravel pits. Fr. 7(1-3). Introduced from Europe.

***A. gigantea** Roth Redtop
(**A. stolonifera** var. **major** (Gaudin) Farw.)
Common. In moist soil in pastures, meadows, waste areas. Fr. 7-8.

A. perennans (Walt.) Tuckerm. Upland Bent
Frequent. On boggy shores, moist woods. Fr. 9.

A. scabra Willd. var. **scabra** Hairgrass
(**A. hiemalis** (Walt.) BSP. var. **tenuis** (Tuckerm.) Gleason)
Frequent. In moist soil on peaty pond shores, wet sandy soil. Fr. 7-8.

***A. spica-venti** L. Wind-spike
(**Apera spica-venti** (L.) Beauv.)
Uncommon. Gravel edge of parking lot and edge of unused gravel road, Moran Wildlife Refuge. Fr. 7. Windsor. First discovered in the county by the author in 1993. From Europe.

A. stolonifera L. Creeping Bent
(**A. alba** var. **palustris** (Huds.) Pers.)
Frequent. In wet meadows, banks of rivers and ponds, wooded swamps. Fr. 7(2-4).

ALOPECURUS

A. aequalis Sobol Short-awned Foxtail
Occasional. In shallow water in shaded pools in fens, ponds. Fr. 6(1-2).

***A. pratensis** L. Meadow Foxtail
Frequent. In wet meadows. Fr. 6(1-2). Introduced from Europe.

ANDROPOGON

A. gerardii Vitm. Big Bluestem
Occasional. On sandy open soil. Sandy floodplain along Deerfield River. Florida, Monterey, Sheffield, Stockbridge, Williamstown. Fr. 7(4)-8(4).

ANTHOXANTHUM

***A. odoratum** L. Sweet Vernal Grass
Common. In dry fields, pastures, roadsides, waste areas. Fr. 5(3-4). Introduced from Europe.

ARISTIDA

A.. dichotomum Michx. Poverty Grass
Occasional. In dry sandy soil, sparse fields and roadsides. Fr. 8(4)-9. Lenox, Williamstown.

***A. necopina** Shinners Three-awn
(**A. longispica** Poir. var. **geniculata** (Raf.) Fern.)
Uncommon. In moist sandy or gravelly soil along railroad tracks. Fr. 9(2-4). First found by the author in 1989.

ARRHENATHERUM

***A. elatius**(L.) Beauv. var. **elatius** Tall Oat-grass
Frequent. On roadsides, fields. Fr. 6(1-4).

AVENA

***A. fatua** L. Oats
Occasional. In waste areas, roadsides, railroad tracks. Fr. 7.

***A. sativa** L Cultivated Oats
Occasional. In waste areas, roadsides. Fr. 7.

BRACHYELYTRUM

B. erectum (Schreb.) Beauv. Long-awned Wood Grass
Common. In openings in moist and/or rich woods. Fr. 7(2)-8(3).

BRIZA

***B. media** L. Quaking Grass
Occasional. In wet calcareous meadows, fields, moist roadsides. Fr. 7(1). Great Barrington, Lee, Washington. West Stockbridge. Introduced from Europe.

BROMUS

B. altissimus Pursh Canada Brome
(Including **B. latiglumis** (Shear) Hitchc.); **B. purgans** of authors, not L.)
Frequent. In moist alluvial soil. Fr. 8-9.

B. ciliatus L. Fringed Brome
Frequent, in marshes, wet meadows. Fr. 7(4)-8.

***B. commutatus** Schrad. Hairy Chess
Uncommon. In waste places. Fr. 6(3). Stockbridge: Hoffmann 1914 (NEBC). Introduced from Europe.

***B. hordeaceus** L. ssp. **hordeaceus** Soft Chess
Uncommon. In waste places, fields. Stockbridge: Hoffmann 1914 (NEBC). Introduced from Europe.

***B. inermis** Leyss. Smooth Brome
Common. On roadsides, fields. Fr. 7-8. Introduced from Europe.

B. kalmii Gray Wild Brome
Occasional. In calcareous wet meadows. Fr. 7-9(3). Southern half of county.

B. pubescens Muhl. ex Willd Canada Brome
Occasional to frequent. In rich rocky woods. Fr. 6(3)-7.

***B. secalinus** L Cheat
Uncommon, in waste areas, railroad tracks. Fr. 7(1-3). Pittsfield. Introduced from Europe.

***B. tectorum** L. Downy Chess
Occasional. Waste places, dry open disturbed areas. Fr. 6(2)-7(3). Introduced from Europe.

CALAMAGROSTIS

C. canadensis (Michx.) Beauv. var. **canadensis** Bluejoint
Common. In wet meadows, marshes, beaver meadows. Fr. 7(1)-8(3).

CENCHRUS

C. longispinus (Hack.) Fern. Sandbur
Uncommon. In dry sandy soil in dry pastures and fields. Sheffield.

CINNA

C. arundinacea L. Broad Woodreed
Frequent. In wooded swamps, usually somewhat calcareous. Fr. 8(2)-9(2).

C. latifolia (Trev. ex Goepp.) Griseb. Drooping Woodreed
Frequent. Cool woods in higher elevations. Fr. 8(2-4).

CYNOSURUS

***C. cristatus** L. Crested Dogtail
Uncommon. In cultivated meadows. Fr. 6(2)-7. Stockbridge: Hoffmann 1917 (NEBC). Introduced from Europe.

DACTYLIS

***D. glomerata** L. Orchard Grass
Common. In fields, roadsides. Fr. 6(1-2). Introduced from Europe.

DANTHONIA

D. compressa Aust. Woodland Oatgrass
Frequent. In clearings in woods, along wood roads. Fr. 7(1-4).

D. spicata (L.) Beauv. ex Roemer & Schultes Poverty Grass
Common. Dry sterile soil in fields. Fr. 6(4)-7.

DESCHAMPSIA

***D. cespitosa** (L.) Beauv. ssp. **parviflora** (Thuill.) Jarmolenko & Soo Tufted Hairgrass
(**D. cespitosa** (L.) Beauv. var. **parviflora** (Thuill.) Coss. & Germ.)
Occasional. In moist soil at edges of woods, roadsides. Fr. 7(4)-8. Stockbridge, Washington, Williamstown. Introduced from Europe.

D. flexuosa (L.) Trin. var. **flexuosa** Common Hairgrass
Frequent. On exposed rocky ledges and ridgetops. Fr. 6(4)-7.

DICHANTHELIUM

The nomenclature of this genus, segregated from **Panicum**, is based on Gould & Clark (1978) with the realignments of Freckmann (1981) used to reinstate certain species. Freckmann's treatment of the **D. acuminatum** complex is also used, except for the inclusion of **D. acuminatum** var. **implicatum**. An effort was made to use the nomenclature that describes most closely those species found in the county.

D. acuminatum (Sw.) Gould & Clark var. **fasciculatum** (Torrey) Freckmann Fascicled Panic Grass
(**P. lanuginosum** Ell. var. **fasciculatum** (Torrey) Fern.)
Common. In dry fields, exposed ledges, sandy soil, peaty soil. Often forming mats. Primary fr. 7(2), secondary fr. 7(4)-9.

D. acuminatum (Sw.) Gould & Clark var. **lindheimeri** (Nash) Gould & Clark Lindheimer's Panic Grass
(**Panicum lanuginosum** (Ell.) Gould var. **septentrionale** Fern.)
Uncommon. In sandy or rocky roadsides, wet woods. Primary fr. 7(1-2). Mt. Washington: Hoffmann 1914 (NEBC).

D. boreale (Nash) Freckmann Northern Panic Grass
(**Panicum boreale** Nash)
Occasional. Wet meadows, moist woods. Primary fr. 6(3). Mt. Washington: Hoffmann 1920 (NEBC). Most common on Berkshire Plateau.

D. clandestinum (L.) Gould Riverside Panic Grass
(**Panicum clandestinum** L.)
Common. On open sandy soil along roads, disturbed areas. Primary fr. 7(3).

D. columbianum (Scribner) Freckmann Downy Panic Grass
(**Panicum columbianum** Scribner)
Frequent. In open dry woods. Primary fr. 6(3)-7(4).

D. commutatum (Schultes) Gould
(**Panicum commutatum** Schultes)
Occasional. Rocky woods, sandy floodplains. Primary fr. 6(3-4). Florida.

D. depauperatum (Muhl.) Gould Depauperate Panic Grass
(**Panicum depauperatum** Muhl.)
Common. Dry sterile soil in fields, open woods. Primary fr. 7(1-2).

D. dichotomum (L.) Gould Forked Panic Grass
(**Panicum dichotomum** L.)
Frequent. Open dry rich woods. Primary fr. 7(1-2).

D. latifolium (L.) Gould & Clark Broadleaf Panic Grass
(**Panicum latifolium** L.)
Common. Rich moist or dry woods and clearings. Primary fr. 6(4)-7(2).

D. linearifolium (Scribner) Gould Linearleaf Panic Grass
(**Panicum linearifolium** Scribner)
Occasional. Open dry woods, fields, ledges. Primary fr. 6(2)-7(2).

D. meridionale (Ashe) Freckmann Pondshore Panic Grass
(**Panicum meridionale** Ashe)
Uncommon. Dry open sandy soil on sandplain. Primary fr. 7(1-2). Sheffield.

D. oligosanthes (Schultes) Gould var.**scribnerianum** (Nash) Gould Scribner's Panic Grass
(**Panicum oligosanthes** Schultes var. **scribnerianum** (Nash) Fern.

Uncommon. Dry sandy soil in grassy sandplain. Primary fr. 7(1-2). Sheffield. First discovered in the county by B. A. Sorrie in 1989.

D. sphaerocarpon (Ell.) Gould Round-fruited Panic Grass
(**Panicum sphaerocarpon** Ell.)
Uncommon,. Edge of dry sandy open area with calcareous outcrops, dry calcareous rubble. Primary fr. 7. Lanesboro. Sheffield.

D. xanthophysum (Gray) Freckmann Yellow Panic Grass
(**Panicum xanthophysum** Gray)
Occasional. Dry rich woods and clearings. Primary fr. 7. New Marlboro. Sheffield: Churchill 1920 (NEBC).

DIGITARIA

D. filiformis (L.) Koel Slender Crabgrass
Uncommon. Sandy fields. Fr. 8(3)-9. Sheffield: Hoffmann 1912 (NEBC).

***D. ischaemum** (Schreb.) Muhl. Smooth Crabgrass
Frequent. Sandy waste areas. Fr. 8(3)-9. Introduced from Europe.

***D. sanguinalis** (L.) Scop. Tall Crabgrass
Common. Waste areas, roadsides, lawns, a garden weed. Fr. 8(3)-9. Introduced from Europe.

ECHINOCHLOA

***E. crusgalli** (L.) Beauv. var. **crusgalli** Barnyard Grass
Common. Waste areas, marshes. Fr. 9-10(1). Introduced from Europe.

***E. crusgalli** (L.) var. **frumentacea** (Roxb.) W. Wight
Billion Dollar Grass
Uncommon. Waste areas, fields. Fr. 8(1)-9(3). Sheffield: Hoffmann 1919 (NEBC). Introduced from Asia.

E. muricata (Beauv.) Fern. var. **muricata** Cockspur Grass
Common. Pond shores, river banks, marshes. Fr. 8(2)-9.

E. muricata (Beauv.) Fern. var. **microstachya** (Wieg.) Fern. & Grisc. Cockspur Grass
Occasional. Moist soil, river banks, roadsides. Fr. 8(2)-9(2). Williamstown.

ELEUSINE

***E. indica** (L.) Gaertn Goosegrass
Uncommon. Gravelly roadside. Great Barrington, Sheffield, Stockbridge. First discovered in the county by B. A. Sorrie in 1989.

ELYMUS

E. canadensis L. var. **canadensis** Canada Wild Rye
Occasional. Alluvial soil on river banks. Fr. 8-9.

E. riparius Wieg. River Wild Rye
Common. Alluvial soil along rivers and streams, occasional in moist woods. Fr. 8-9(2).

E. villosus Muhl. ex Willd. **T** Hairy Wild Rye
Rare. In rich dryish alluvial soil on river banks. 7(2)-8(2). Sheffield. Great Barrington.

E. virginicus L. var. **virginicus** Virginia Wild Rye
Frequent. Moist or wet soil on river and stream banks, in marshes, on moist banks. Fr. 7(4)-9(2).

E. wiegandii Fern. Wiegand's Wild Rye
(**E. canadensis** L. var. **wiegandii** (Fern.) Bowden)
Frequent. Rich alluvial soil on river banks. Fr. 7(4)-8.

ERAGROSTIS

E. capillaris (L.) Nees Lace Lovegrass
Uncommon. Sandy soil. 7(3)-8. Sheffield: Hoffmann 1914 (NEBC).

***E. cilianensis** (All.) Mosher Stinkgrass
Frequent. Sandy waste areas and roadsides. Fr. 8(2)-10(4). Introduced from Europe.

E. frankii C. A. Mey. ex Steud. **SC** Frank's Lovegrass
Rare. On silt bars in Housatonic River, also on moist roadside. Alford, Great Barrington, Lee, Richmond, Sheffield Stockbridge, Williamstown.

E. hypnoides (Lam.) BSP. Creeping Lovegrass
Occasional. Silt or mud shores of rivers. Fr. 8(2)-9(3).

E. pectinacea (Michx.) Nees Tufted Lovegrass
Common. Waste areas, railroad tracks, roadsides, silt bars. Fr. 8(1)-9.

***E. pilosa** (L.) Beauv. var. **pilosa** India Lovegrass
(**E. multicaulis** Steud.; **E. peregrina** Wieg.)
Common. Waste areas, roadsides, railroad tracks. Fr. 8(3)-9. Introduced from Europe.

E. spectabilis (Pursh) Steud. Purple Lovegrass
Frequent. Sandy soil in fields, roadsides. Fr. 8(4)-9.

FESTUCA

***F. arundinacea** Schreber Tall Fescue
(**F. elatior** L. in part)
Common. Fields, on roadsides. Fr. 7. Introduced from Europe.

F. obtusa Biehl. Nodding Fescue
Common. In rich woods, moist woods. Fr. 7.

***F. ovina** L. Sheep Fescue
Common. Dry fields, roadsides. Fr. 6(4). Introduced from Europe.

***F. pratensis** Hudson Meadow Fescue
(**F. elatior** L. in part)
Common. Fields, roadsides. Introduced from Europe.

F. rubra L. Red Fescue
(Including varieties)
Common. Dry fields, roadsides. Fr. 6(2). Native and introduced from Europe.

***F. tenuifolia** Sibth. Hair Fescue
(**F. filiformis** Pourret; **F. capillata** Lam.)
Frequent. Dry sterile fields and sandy roadsides, river bank. Fr. 6. Introduced from Europe.

GLYCERIA

G. acutiflora Torr. Sharpscale Mannagrass
Uncommon. Shallow water or muddy margins of ponds, floodplains. Fr. 7-8. Great Barrington, Sheffield. Stockbridge: Hoffmann 1919 (NEBC).

G. borealis (Nash) Batch. Northern Mannagrass, Float-grass
Frequent. Shallow water at edges of ponds, streams, in wet meadows. Fr. 6(2)-8(1).

G. canadensis (Michx.) Trin. Rattlesnake Grass
Common. Marshes, wet woods, wet meadows. Fr. 6(4)-8(2).

G. grandis S. Wats. Reed Meadow Grass
Frequent. Marshes, edges of wet woods. Fr. 6(3)-7(2).

G. melicaria (Michx.) F. T. Hubbard Slender Mannagrass
Frequent. Wet woods, in shaded swamps and along streams. Fr. 8.

G. septentrionalis Hitchc. Eastern Mannagrass
Occasional. Mud of basin pools, often temporary, in woods. Fr. 7(2)-8. Hancock, Mt. Washington, New Marlboro, Sheffield.

G. striata (Lam.) Hitchc. Fowl Mannagrass
Common. Wet meadows, ditches, marshes, margins of ponds. Fr. 5(4)-6(3).

HIEROCHLOE

H. odorata (L.) Beauv. Indian Sweetgrass
Occasional. Wet meadows, graminoid fens. Fr. 5(4)-6(2). Lenox, Savoy, Sheffield.

HOLCUS

***H. lanatus** L. Velvet Grass
Occasional. Moist fields. Fr. 6(4)-7(4). Stockbridge. New Marlboro, Williamstown. Introduced from Europe.

HORDEUM

***H. jubatum** L. Squirrel-tail Grass
Occasional. Railroad tracks, roadsides, waste areas. Fr. 6(4)-7. Hinsdale, Lanesboro, New Marlboro, Windsor.

***H. vulgare** L. Barley
Occasional. Roadsides, railroad tracks. Fr. 7(2). Pittsfield, Williamstown. Great Barrington: Hoffmann 1919 (NEBC). Introduced from Europe.

HYSTRIX

H. patula Moench Bottlebrush Grass
(**Elymus hystrix** L. var. **hystrix)**
Frequent. Moist rich woods. Fr. 7-8.

H. patula Moench var. **bigeloviana** (Fern.) Deam
(**Elymus hystrix** L. var. **bigeloviana** (Fern.) Bowden
Occasional. Rich woods and thickets. Fr. 7-8.

LEERSIA

L. oryzoides (L.) Sw. Rice Cutgrass
Common. Marshes, wet meadows. Fr. 8.

L. virginica Willd. Whitegrass
Frequent. Moist rich woods, moist soil along streams. Fr. 8.

LEPTOLOMA

***L. cognatum** (Schultes) Chase Witch Crabgrass
Uncommon. Dry sandy soil in open fields. Fr. 8(3)-9(3). Lee, Sheffield. Probably adventive from farther west. This grass has been spreading rapidly in New England (Zika, 1990). First discovered in the county by B. A. Sorrie in 1991.

LOLIUM

***L. multiflorum** Lam. Italian Ryegrass
Occasional. Disturbed areas, seeded areas Fr. 7-9(2). Introduced from Europe.

***L. perenne** L. Perennial Ryegrass
Common. Roadsides. Fr. 6(2)-8(3). Introduced from Europe.

MILIUM

M. effusum L. **T** — Woodland Millet
Rare. Rich moist soil, woods at higher elevations. Fr. 6(3)-8(4). Frequent on Mt. Greylock. Adams, Dalton, Hinsdale, New Ashford, Sandisfield, Savoy, West Stockbridge, Williamstown.

MOLINIA

***M. caerulea** (L.) Moench. — Moorgrass
Uncommon. Edge of calcareous wooded swamp. Fl. 7(2). First discovered by the author in 1991.

MUHLENBERGIA

M. frondosa (Poir.) Fern. — Wirestem Muhly
Common. Shaded river banks. Fr. 8(3)-9(2).

M. glomerata(Willd.) Trin. — Spike Muhly
Frequent. Calcareous wet meadows and seeps. Fr. 8-9(1).

M. mexicana (L.) Trin. — Muhly
Occasional. Edges of woods, moist soil along railroads. Fr. 9 (1-3).

M. schreberi Gmel. — Nimble-will
Common. Moist soil in disturbed areas. Fr. 8(4)-9(3). Often a annoying weed in gardens.

M. sylvatica (Torr.) Torr. ex Gray — Woodland Dropseed
Occasional. Dry to moist rich soil in woods and rocky flood-plain. Fr. 8(3)-9(2).

M. tenuiflora (Willd.) BSP. — Woodland Dropseed
Occasional. Dry rich soil on rocky wooded slopes. Fr. 8(3)-9(2). Mostly in southern half of county.

M. uniflora (Muhl.) Fern. — Pondshore Muhly
Uncommon. In wet peaty soil on margin of ponds, clearings. Fr. 7(4)-8(4). Becket, Peru, Savoy.

ORYZOPSIS

O. asperifolia Michx. — Rough Mountain-rice
Common. Dry woods, occasional in rich woods. Fr. 6(2-3).

O. pungens (Torr. ex Spreng.) Hitchc. — Slender Mountain-rice
Uncommon. On quartzite summit of Monument Mt. Fr. 6(1-3). Great Barrington Hoffmann 1915 (NEBC).

O. racemosa (Sm.) Ricker — Black Mountain-rice
Frequent. Rich moist soil in woods, shaded calcareous ledges. Fr. 8.

PANICUM

P. capillare L. — Witchgrass
(Including **P. capillare** L. var. **occidentale** Rydb.)
Common. Dry soil in waste areas, roadsides, railroads. Fr. 7(4)-9(2).

P. dichotomoflorum Michx. var. **dichotomoflorum** — Smooth Panic Grass
Common. Waste areas, railroads, gravel roadsides. Fr. 9(1-4). First discovered in the county by H. E Ahles in 1967.

P. gattingeri Nash **SC** — Gattinger's Panic Grass
Rare. On sandy or gravelly roadsides. Fr. 8(3)-9(2). North Adams, Stockbridge. First discovered in the county by H. E. Ahles in 1979.

***P. miliaceum** L. — Broomcorn Millet
Occasional. Waste areas, railroads. Fr. 8-9. Washington. Great Barrington: Hoffmann 1920 (NEBC). Introduced from Europe.

P. philadelphicum Bernh. ex Trin. **SC** — Philadelphia Panic Grass
(Including **P. philadelphicum** var. **tuckermanii** (Fern.) Steyerm. & Schm.)
Rare. On sandy or peaty pond shores. Fr. 8(2-4). New Marlboro: Hoffmann 1920 (NEBC).

P. rigidulum Bosc ex Nees var. **rigidulum** — Panic Grass
(**P. agrostoides** Spreng.)
Uncommon. Marshes along streams, sandy pond shores. Fr. 9-10(2). Otis, Sheffield.

***P. virgatum** L. var. **virgatum** — Switch Grass
Uncommon. Dry soil near roadside, railroad. Fr. 7(2-4). Sheffield, Williamstown. Probably spread from plantings along highway. First discovered in the county by the author in 1989.

PASPALUM

P. setaceum Michx. var. **muhlenbergii** (Nash) D. Banks — Muhlenberg's Paspalum
(**P. ciliatifolium** Michx. var.**muhlenbergii** (Nash) Fern.)
Frequent. Dry sandy, usually somewhat calcareous fields and pastures. Fr. 8(2)-9(4).

PHALARIS

P. arundinacea L. — Reed Canary-grass
Common. Wet soil in meadows, roadsides, marshes, flood-plains. Fr. 6(3)-7. Increasingly widespread. Both native and introduced from Europe, the introduced plants possibly being aggressive and mat-forming (Dore & McNeill 1980).

PHLEUM

***P. pratense** L. Timothy
Common. Fields, pastures, roadsides. Fr. 6(3)-8(3). Introduced from Europe.

PHRAGMITES

***P. australis** (Cav.) Trin. ex Steud. Common Reedgrass
(**P. communis** (L.) Trin.)
Common. Moist to wet disturbed soil, roadsides, marshes, wet meadows, bogs and fens. Fr. 8(3)-9(3). Hoffmann (1922) noted only two stations, while at present it is in every town, abundant and spreading, threatening valuable and rare habitats. Usually where soil has been disturbed or road salt use is heavy, as along roads, gas lines, cable lines. It reproduces mainly vegetatively (Davis & Briggs, 1986), and fragments of rhizomes are spread by construction machinery. Native to brackish marshes.

POA

P. alsodes Gray Woodland Bluegrass
Common. Moist rich soil, in woods, along streams and shaded swamps. Fr. 5(3-4).

***P. annua** L. Speargrass
Common. Lawns, waste areas, roadsides. Fr. 6-9. Introduced from Eurasia.

***P. compressa** L. Canada Bluegrass, Wiregrass
Common. Dry ledges, waste areas, roadsides. Fr. 6(2)-8(2). Introduced from Eurasia.

P. languida Hitchc. **E** Drooping Speargrass
Rare. Rich calcareous dry rocky soil. Open oak-hickory-hop hornbeam woodland. Fr. 6(1-2). Egremont, Great Barrington.
***P. nemoralis** L. Wood Bluegrass
Occasional. Roadsides, banks of brooks. Fr. 8(3)-9(1). Introduced from Europe.

P. palustris L. Fowl Meadowgrass
Common. Wet meadows, river banks. Fr. 7(1)-8(2).

***P. pratensis** L. Kentucky Bluegrass
Common. Fields, lawns, clearings. Fr. 6(2)-7(4). Ours probably introduced from Europe.

P. saltuensis Fern. & Wieg. Woodland Bluegrass
Frequent. Dry open woods. Fr. 6(1-3).

***P. trivialis** L Rough Bluegrass
Occasional. River banks, swampy woods. Fr. 6(3)-7(3). Sheffield, Egremont. Introduced from Europe.

PUCCINELLIA

***P. distans** (Jacq.) Parl. Alkali-grass
Uncommon. In crushed limestone, roadsides that are heavily salted. Fr. 6(1-4). Lee, Richmond, Williamstown. Introduced from Europe.

SCHIZACHNE

S. purpurascens (Torr.) Sw. Purple False Melic
Frequent. Moist rich woods, bases of calcareous ledges. Fr. 6(1-3).

SCHIZACHYRIUM

S. scoparium (Michx.) Nash ssp. **scoparium** Little Bluestem
(**Andropogon scoparius** Michx.)
Common. Dry sandy soil, in fields, railroads, roadsides, open ridgetops. Fr. 8(3)-9.

SECALE

***S. cereale** L
Rye
Occasional. Roadsides, railroads. Fr. 6(3)-8(3). Pittsfield: Hoffmann 1902 (NEBC). Used in seeding mixtures.

SETARIA

***S. faberi** Herrm. Giant Foxtail
Frequent. Sandy or gravelly waste areas, roadsides, railroads. Fr. 9. Lee, North Adams. Great Barrington. First discovered in the county in 1988 by the author. Introduced from Asia.

***S. glauca** (L.) Beauv. Foxtail, Pigeongrass
Common and abundant. Roadsides, waste areas. Fr. 9. Introduced from Eurasia.

***S. italica** (L.) Beauv. Foxtail Millet
Uncommon. Waste areas, near farms. Fr. 7-8. Becket: Hoffmann 1904 (NEBC). Introduced from Europe.

***S. viridis** (L.) Beauv. Bottlegrass
Occasional. Roadsides, waste areas, cultivated fields. Fr. 7(2)-9(1). Introduced from Europe.

SORGHASTRUM

S. nutans (L.) Nash Indian Grass
Occasional. Dry sandy fields, roadsides. Pittsfield, Sheffield. Great Barrington. Mostly in southern half of county.

SPARTINA

***S. pectinata** Link Freshwater Cordgrass
Uncommon. Moist soil along railroad. Fr. 9(3). Lee. Richmond. First discovered in the county by H. E. Ahles in 1967.

Introduced from farther east or west.

SPHENOPHOLIS

S. nitida (Biehl.) Scribn. **T** Shining Wedgegrass
Rare. Rich dry soil. In open oak-hickory- hop hornbeam woodland. Fr. 5(3)-6. Great Barrington, Sheffield, Stockbridge, New Marlboro.

S. obtusata (Michx.) Scribn. var. **major** (Torr.) Erdman
(**S. intermedia** (Rydb.) Rydb.) Slender Wedgegrass
Occasional. Shaded calcareous ledges and rocks along rivers, and in calcareous marshes. Fr. 7(3-4). In shaded situations, the species has a lax, drooping habit, while specimens in open marsh are erect and taller.

SPOROBOLUS

*__S. asper__ (Michx.) Kunth Rough Dropseed
Uncommon. Dry sandy soil along road and railroad. Fr. 9(2-4). Stockbridge. Great Barrington. First discovered in the county by the author in 1989. Probably introduced from farther west.

*__S. cryptandrus__ (Torr.) Gray Sand Dropseed
Uncommon. Dry sandy roadside. Fr. 8(4). Sheffield. First discovered in the county by B. A. Sorrie and the author. Introduced from farther south and west. Spreading in Vermont (Zika, 1990).

S. neglectus Nash **E** Small Dropseed
Rare. In dry calcareous soil, sandy or limy rubble. Fr. 9(1). Sheffield.

S. vaginiflorus (Torr. ex Gray) Wood Poverty Grass
Common. Waste areas, roadsides. Fr. 8(3)-9(1).

TORREYOCHLOA

T. fernaldii (Hitchc.) Voss Slender Mannagrass
(**Glyceria fernaldii** (Hitchc.) St. John)
Uncommon. River swales, shallow water. Fr. 6(2)-7(3). Lenox, Sheffield: Churchill 1919 (NEBC).

T. pallida (Torrey) Church Pale Mannagrass
(**Glyceria pallida** (Torr.) Trin.)
Uncommon. In shallow water of marshes and pools. Fr. 6(2)-7(3). New Marlboro, Sandisfield.

TRIDENS

T. flavus (L.) Hitchc. Purpletop
(**Triodia flava** (L.) Smyth)
Uncommon. Dry sandy field. Fr. 8(4)-9. Cheshire. Listed by Dewey (in Field, 1829). Spreading in New England (Zika, 1990).

TRISETUM

T. triflorum (Bigel.) Löve & Löve ssp. **molle** (Michx.) Löve & Löve Spiked False-oats
(**T. spicatum** (L.) Richter)
Rare. On shore of Deerfield River. Fr. 6-7(3). Florida.

ZIZANIA

*__Z. palustris__ L. var. **palustris** Wild Rice
(**Z. aquatica** L. var. **angustifolia** Hitchc.)
Uncommon. In shallow water in acidic ponds at higher elevations (420-640 m), also more alkaline marshes. Fr. 7. Hancock, Pittsfield, Sandisfield. Possibly spread from plantings for wild fowl. Not listed by Hoffmann.

SPARGANIACEAE

SPARGANIUM

S. americanum Nutt. Bur-reed
Common. In shallow water in marshes, along streams. Fr. 8(4)-9(1-3).

S. angustifolium Michx. Narrow-leaved Bur-reed
(**S. emersum** Rehm. var. **angustifolium** (Michx.) Taylor & MacBryde)
Occasional. In cool acidic ponds, on Berkshire Plateau and Taconic Range. Fr. 8(1). Mt. Washington, New Marlboro, Otis. Savoy.

S. emersum Rehm. Bur-reed
(**S. chlorocarpum** Rydb.)
Common. In shallow water, on pond shores, in marshes, in streams and rivers, often in current where leaves become elongated and floating. Fr. 9(1).

S. eurycarpum Engelm. Large-fruited Bur-reed
Frequent. Rich deep muck in calcareous marshes. Fr. 8(2)-9(1). In the valley region.

S. fluctuans (Morong) Robins. Floating Bur-reed
Occasional. In cool acidic ponds, deep water. Fr. 8(2-3). Otis, Washington. Savoy.

S. natans L. **E** Northern Bur-reed
(**S. minimum** (Hartm.) Fries)
Rare. In peaty pools adjacent to central pond of basin fen. Fr. 7(2-3). Stockbridge: Sorrie et al. 2456 (pers. herb.).

TYPHACEAE

TYPHA

T. angustifolia L. Narrow-leaf Cattail

Occasional. Calcareous, alkaline or brackish wet soil as along salted roads, in shallow water, marshes. Fl. 7(1). Pittsfield, Richmond, Stockbridge, Williamstown.

T. Xglauca Godr. Hybrid Cattail
Occasional to frequent. In water at edges of marshes, streams and ponds, usually in calcareous soil. Fl. 6(3). Pittsfield, Richmond, West Stockbridge, Stockbridge.

T. latifolia L. Common Cattail
Common and abundant. In shallow water in marshes, ditches, along streams and in wet soil. Fl. 6(3).

PONTEDERIACEAE

PONTEDERIA

P. cordata L. Pickerel-Weed
Frequent. Marshy margins of ponds, bog ponds, slow rivers. Fl. 7-8.

HETERANTHERA

H. dubia (Jacq.) MacM. Water Star-grass
(**Zosterella dubia** (Jacq.) Small)
Occasional. In shallow calcareous water, on muddy shores of slow rivers, ponds. Fl. 8(2) rarely, usually when water levels are low. and the foliage reaches the surface. Cheshire, Great Barrington, Lanesboro, Monterey, Pittsfield, Richmond, Sheffield, West Stockbridge, Williamstown. Plate 8B.

LILIACEAE

ALLIUM

A. canadense L. Wild Garlic
Occasional. Rich alluvial shaded soil. Fl. 6(3-4). Egremont, Sheffield, Williamstown. Possibly more common, as not easily noticed.

A. tricoccum Ait. Wild Leek
Common. Moist rich woods, often forming large patches. A good indicator species for rich sites. Lvs. 4(3), Fl. 6(4)-7(2).

ASPARAGUS

***A. officinalis** L. Asparagus
Frequent. Roadsides, edges of woods, fields. Fl. 6(3). Introduced from Europe.

CHAMAELIRIUM

C. luteum (L.) Gray **E** Fairy Wand, Devil's Bit
Rare. In open, dry, rich woods with calcareous rock outcrops. Fl. 6(3)-7(1). Egremont, Sheffield. Great Barrington. In the period 1902-1920, flowering plants were collected, and Hoffmann (1922) noted that it was "occasional in the southern part of the valley." At present, few inflorescences are seen, and vegetative plants are few and scattered. Possibly this is due to succession to more mature forest or invasion of brushy growth and aliens after logging. The species is dioecious, and recovery is difficult. Plate 8C.

CLINTONIA

C. borealis (Ait.) Raf. Bluebead Lily, Clintonia
Common. Moist cool upland woods, both deciduous and spruce-fir. Fl. 5(3)-6(2).

CONVALLARIA

***C. majalis** L. Lily-of-the-Valley
Occasional. Woods along roads. Fl. 5(2). Introduced from Europe.

ERYTHRONIUM

E. americanum Ker Trout Lily, Dog-tooth Violet
Common. Open moist woods and clearings. Fl. 4(3)-6(2).

HEMEROCALLIS

***H. fulva** L. Common Day Lily
Common. Roadsides, edge of woods near old house sites. Fl. 6(2)-8(2). Introduced from Eurasia.

HYPOXIS

H. hirsuta (L.) Cov. Yellow Stargrass
Uncommon. Open woods and fields, ledges. Fl. 6(1-4). Mt. Washington: Churchill 1919 (NEBC). Hoffmann (1922) termed the species frequent but it has not been seen recently.

LILIUM

L. canadense L. Canada Lily
Frequent. Open or shaded marshes, wet meadows, wet borders of streams. Fl. 6(4)-7(3).

L. philadelphicum L. Wood Lily
Occasional. Dry acidic soil in open woods, powerlines and clearings. Fl. 7. Great Barrington, Savoy, Sheffield, Williamstown.

MAIANTHEMUM

M. canadense Desf. Canada Mayflower
Common and abundant. Forms large patches in both deciduous and conifer woods. Fl. 6(1-3).

MEDEOLA

M. virginiana L. Indian Cucumber-root

Common. Open woods and along trails. Fl. 5(4)-6(3).

POLYGONATUM

P. biflorum (Walt.) Ell. var.**commutatum** (Schultes f.) Morong (**P. canaliculatum** of authors, not (Muhl.) Pursh) Giant Solomon's Seal
Occasional. Alluvial soil in floodplain forests. New Marlboro, Sheffield, Williamstown. Fl. 6(2). (See Voss, 1978).

P. pubescens (Willd.) Pursh Solomon's Seal
Common. Moist to dry woods. Fl. 5(2-4).

SMILACINA

S. racemosa (L.) Desf. False Solomon's Seal
(**Maianthemum racemosum** (L.) Link ssp. **racemosum**)
Common. Edges of woods, roadsides, in clearings. Fl. 5(4)-6(2).

S. stellata (L.) Desf. Starry False Solomon's Seal
(**Maianthemum stellatum** (L.) Link)
Frequent. Alluvial soil along rivers, seeps and on high open ledges. Fl. 5(3)-6(1).

S. trifolia (L.) Desf. Three-leaved False Solomon's Seal
(**Maianthemum trifolium** (L.) Sloboda)
Uncommon. In bogs on sphagnum mat in shrub or forested zones. Fl. 6(1-3). New Marlboro, Otis, Peru,Washington.

STREPTOPUS

S. amplexifolius (L.) DC. var. **amplexifolius** White Mandarin
Occasional. In cold wet ravines and ledges, more common at higher elevations. Fl. 5(4)-6(3). Adams, Florida, Mt. Washington, Sandisfield, Windsor.

S. roseus Michx. var. **perspectus** Fassett Rosy Twisted-stalk
Frequent. Moist cool woods. Fl. 5(4)-6(3).

TRILLIUM

T. cernuum L. var. **cernuum** Nodding Trillium
Uncommon. Rich alluvial soil in river floodplains and along streams. Fl. 5(1-4). Richmond, Sheffield.

T. erectum L. Wake-robin, Purple Trillium
Common. Moist rich woods, moist woods. Fl. 4(4)-5(2). A pale cream color form is occasionally seen.

T. undulatum Willd. Painted Trillium
Frequent. Moist cool acidic woods, often under conifers. Fl. 5(2-4).

UVULARIA

U. grandiflora Sm. Large-flowered Bellwort
Frequent. Moist rich woods, rich soil near calcareous outcrops. Fl. 4(4)-5(3). Most towns in the central valley region.

U. sessilifolia L. Wild-oats, Bellwort
Common. Dry or moist woods, clearings. Fl. 4(3)-5(2).

U. perfoliata L. Perfoliate Bellwort
Frequent. Dry rich open woods. Fl. 5(2-4). More common in southern half of county.

VERATRUM

V. viride Ait. False Hellebore, Indian-poke
Common. Moist soil along streams, marshes, in wet depressions, wet meadows. Fl. 6(3).

IRIDACEAE

IRIS

***I. pseudacorus** L. Yellow Iris
Frequent. Marshes, wet meadows, borders of ponds. Fl. 6(1-3). May crowd out native species. Introduced from Europe. Not noted by Hoffmann.

I. versicolor L. Blue Flag, Wild Iris
Common. Wet meadows, marshes. Fl. 5(4)-7(2).

***I. siberica** L. Siberian Iris
Uncommon. Waste area. Fl. 6. Pittsfield: Ahles 1973 (MASS). Introduced from Eurasia.

SISYRINCHIUM

S. angustifolium Mill. Stout Blue-eyed Grass
(**S. graminoides** Bickn.)
Frequent. Moist soil in clearings, ditches, wet meadows. Fl. 5(4)-6(4).

S. atlanticum Bickn Eastern Blue-eyed Grass
Frequent. Moist sandy soil, in meadows. Fl. 6(2)-7(1).

S. montanum Greene var. **crebrum** Fern. Meadow Blue-eyed Grass
Frequent. Wet meadows, ditches. Fl. 5(3)-6(1).

S. mucronatum Michx. **T** Slender Blue-eyed Grass
Rare. In calcareous wet meadows. Fl. 5(3)-6(2). Becket, Richmond, Williamstown.

SMILACACEAE

SMILAX
S. herbacea L. Carrion-flower

Common. Moist open woods at border of marshes, swampy woods. Fl. 6(1-2).

S. rotundifolia L. Common Catbrier
Uncommon. Moist woods. Fl. 5(4)-6(2). Sheffield: Walters 1915 (NEBC).

ORCHIDACEAE

ARETHUSA

A. bulbosa L. **H** Arethusa
Historical. In open bogs on sphagnum mat. Fl. 6(1-2). Monterey. Great Barrington: Walters 1915 (NEBC). Hoffmann (1922) noted the species was "occasional". It is apparently very scarce now, possibly because some peat bogs have been destroyed and open mats have become shrubby.

CALOPOGON

C. tuberosus (L.) BSP. Grass Pink
(**C. pulchellus** (Salisb.) R. Br. ex Ait.)
Uncommon. Open mat of bogs and graminoid fens. Fl. 6(2)-7(2). Otis, Sheffield, Stockbridge, West Stockbridge. Plate 8D.

COELOGLOSSUM

C. viride (L.) Hartm. var. **virescens** (Muhl. ex Willd.) Luer (**Habeneria viridis** (L.) R. Br. ssp. **bracteata** (Muhl. ex Willd.) A. Gray) Long-bracted Orchid
Uncommon. Moist rich woods. Fl. 5(3)-6. Florida, Great Barrington, Sheffield, Williamstown.

CORALLORRHIZA

C. maculata (Raf.) Raf. Spotted Coralroot
Occasional to frequent, in moist rich woods. Fl. 7(4)-8(2).

C. odontorhiza (Willd.) Nutt. **SC** Autumn Coralroot
Rare. In dry rich woods. Fl. 8(3). Sheffield. Not listed by Hoffmann, 1922) but listed by Dewey (in Field, 1829). An inconspicuous orchid that is easily missed.

C. trifida Chat. Early Coralroot
Uncommon. Swampy woods, wet banks. Fl. 6(1)-7(2). Great Barrington, Hinsdale (Sorrie sight record 1986). West Stockbridge: Evans, Fernald & Knowlton 1919 (NEBC).

CYPRIPEDIUM

C. acaule Ait Pink Ladyslipper, Moccasin Flower
Frequent. Dry open acidic sunny woods. Fl. 5(2)-6(1).

C. arietinum R. Br. **E** Ram's Head Ladyslipper
Rare, and probably historical in the county. In rich woods. Fl. 5(3-4). Great Barrington: Brown between 1908-11 (NEBC).

C. calceolus L. var. **parviflorum** (Salisb.) Fern. **E**
(**C. parviflorum** Salisb.) Small Yellow Ladyslipper
Rare. In forested fens. Fl. 5(3)-6(1). Hinsdale, Sheffield, Stockbridge, Williamstown. Plate 8E.

C. calceolus L. var. **pubescens** (Willd.) Correll
(**C. pubescens** Willd.) Yellow Ladyslipper
Occasional. Calcareous swamps and moist rich woods. Fl. 5(3-4). Great Barrington, Hancock, Hinsdale, Lanesboro, New Marlboro, Pittsfield, Sheffield , West Stockbridge, Williamstown.

C. reginae Walt. **SC** Showy Ladyslipper
Rare. In forested fens, seepage swamps on slopes, calcareous swamps. Fl. 6(3-4). Most valley region towns. This handsome orchid is always under threat of being picked or browsed by deer.

EPIPACTIS

***E. helleborine** (L.) Crantz Helleborine
Common. Moist woods, roadsides, everywhere. Our only "weed" orchid. Fl. 7(2)-9(1). Introduced from Europe.

GALEARIS

G. spectabilis (L.) Raf. Showy Orchis
(**Orchis spectabilis** L.)
Frequent. Moist rich woods, rich soil below calcareous outcrops. Fl. 5(3)-6(1).

GOODYERA

G. pubescens(Willd.) R. Br. Downy Rattlesnake-plantain
Frequent. Open rich woods, dry woods, moist rich banks. Fl. 7(4)-8(4).

G. repens (L.) R. Br. **E** Dwarf Rattlesnake-plantain
Rare. Under red spruce on Mt. Greylock, under hemlock on serpentine. Fl. 7(20-8(3). Florida, Williamstown.

G. tesselata Lodd. Checkered Rattlesnake-plantain
Occasional. Open moist woods, along brooks, in clearings along trails. Fl. 7(4)-8(3). Adams, Florida, Great Barrington, North Adams, Sandisfield, Sheffield, Stockbridge, Williamstown.

ISOTRIA

I. verticillata (Muhl. ex Willd.) Raf. Large Whorled Pogonia
Occasional. Dry, open south-facing dry oak-blueberry dominat-

ed woods. Fl. 5(3)-6(1). Clarksburg, Great Barrington, North Adams, Sheffield, Williamstown. Plate 8F.

LIPARIS

L. lilifolia (L.) Rich. ex Lindl. Lily-leaved Twayblade
Uncommon. Rich or rocky woods. Fl. 6(2)-7(1). Alford, West Stockbridge, Williamstown. Mt. Washington: Walters 1915 (NEBC).

L. loeselii (L.) Rich. Bog Twayblade, Fen Orchid
Occasional. On mossy tussocks in forested fens. Fl. 6(2)-7(2).

LISTERA

L. cordata (L.) R. Br. **H** Heartleaf Twayblade
Rare and historical. In wet coniferous woods. Fl. 6(2)-7(3). Washington: Boott 1856 (NEBC). Hoffmann did not list this species, perhaps because of the age of the specimen.

MALAXIS

M. brachypoda T White Adder's Mouth
Rare. In bare wet soil in depressions in forested fens, swampy woods. Fl. 6(1)-7(1). Pittsfield, Sheffield, Stockbridge, Windsor, Williamstown.

M. unifolia Michx. Green Adder's Mouth
Uncommon. Dry mixed woods, pastures, young woods, fens. Fl. 7(3)-8(1). Adams, Becket, Hancock, Peru, Pittsfield, Sheffield, Stockbridge.

PLATANTHERA (HABENERIA)

P. Xandrewsii (White) Luer Fringed Orchis
(**P. lacera** X **P. psychodes**)
Uncommon. Wet meadows. Fl. 8(2). Becket. Windsor: Quinlan & Gola 1983 (NEBC).

P. blephariglottis (Willd.) Lindl. White Fringed Orchis
Uncommon. Near shrub border of open mat of bogs, in sphagnum. Fl. 7(2)-8(3). Sheffield, Otis.

P. clavellata (Michx.) Luer Green Woodland Orchis
Frequent. In sphagnum mat of bogs, peaty borders of ponds, wet woods. Fl. 8(1-3).

P. dilatata (Pursh) Lindl. ex Beck **T** Leafy White Orchis
Rare. On margins of bogs, peaty borders of ponds. Fl. 6(3-4). Lanesboro, Great Barrington. Savoy: Quinlan 1981 (NEBC).

P. flava (L.) Lindl. var. **herbiola** (R. Br.) Luer **T**
Pale Green Orchis, Tubercled Orchis
Rare. In calcareous wet meadows, floodplains. Fl. 6(3)-7(3). Florida, Lenox, Sheffield. Hoffmann (1922) noted that this species was "frequent".

P. grandiflora (Bigel.) Lindl. Large Purple-fringed Orchis
(**Habeneria fimbriata** (Ait.) R. Br.)
Uncommon. Cool moist woods, somewhat rich woods. Fl. 7(1)-8(1). Becket, Cheshire, Florida, Hinsdale, Sandisfield, Washington. On Berkshire Plateau and Mt. Greylock Range.

P. hookeri (Torr. ex Gray) Lindl. Hooker's Orchis
Uncommon. Rich dry woods. 6(1)-8(1). Lenox, New Marlboro. Sheffield: Churchill 1920 (NEBC). Peru: Sorrie 1986 sight record. Hoffmann noted this species was "frequent".

P. hyperborea (L.) Lindl. Northern Green Orchis
Common. Wet woods, seepy banks, along streams. Fl. 6(3)-7(4).

P. lacera (Michx.) G. Don Ragged Fringed Orchis
Frequent. Wet meadows, seepy pastures. Fl. 7(2)-8(3).

P. obtusata (Banks ex Pursh) Lindl. **H** Bluntleaf Orchis
Rare. In wet mossy spruce-fir woods or bog forest. Fl. 6(3)-7(3). Hinsdale. Peru: Harger 1879 (NEBC).

P. orbiculata (Pursh) Lindl. var. **macrophylla** (Goldie)Luer
(**Habeneria macrophylla** Goldie) Large Round-leaf Orchis
Uncommon. Rich woods. Fl. 6(3)-8(2). Florida, New Marlboro, Peru, Williamstown.

P. orbiculata (Pursh) Lindl. var. **orbiculata** Roundleaf Orchis
(**Habeneria orbiculata** (Pursh) Torr.)
Uncommon. Moist cool woods. Fl. 7. Florida, Cheshire, Peru, Sandisfield, Savoy, Washington.

P. psycodes (L.) Lindl. Small Purple-fringed Orchis
Frequent. Wet meadows, peaty margins of ponds. Fl. 7(4)-8(4).

POGONIA

P. ophioglossoides (L.) Juss. Rose Pogonia, Snake-mouth
Occasional. In open mat of bogs, on tussocks in graminoid fens, peaty shores. Fl. 6(3)-7(4). Clarksburg, Otis, Savoy, Stockbridge, Sheffield.

SPIRANTHES

S. cernua (L.) Rich. Nodding Ladies'-tresses
Frequent. Wet meadows, calcareous seeps, moist roadsides. Fl. 8(4)-9.

S. lacera (Raf.) Raf. var. **gracilis** (Bigel.) Luer
(**S. gracilis** (Bigel.) Beck) Slender Ladies'-tresses
Uncommon. In dry somewhat calcareous fields. Fl. 7(2)-8(1).

Cheshire, Lanesboro, Sandisfield, Sheffield, West Stockbridge, Williamstown. Noted as "common" by Hoffmann (1922), it is rarely seen now.

S. lucida (H. Eat.) Ames Shining Ladies'-tresses
Frequent. Moist disturbed situations, in wet meadows and pastures, seeps, moist pond margins. Fl. 6(2)-7.

S. ochroleuca (Rydb.) Rydb. Creamy Ladies'-tresses
(**S. cernua** var. **ochroleuca** (Rydb.) Ames)
Uncommon. Dry calcareous soil in open disturbed fields, roadsides. Fl. 9. Becket, Lanesboro, Lenox, Sheffield.

S. romanzoffiana Cham. **E** Hooded Ladies'-tresses
Rare. In open calcareous gravelly seeps, edges of fens. Fl. 8(1-3). Egremont, Hinsdale, Monterey, Peru, Pittsfield.

Non-persistent Species

The following taxa are represented by specimens in herbaria, but are considered not to be naturalized members of the flora. Hoffmann (1922) included most of these in his main catalogue, but also listed them in a "fugitive species" category, recognizing their ephemeral status. Many were found only on wool dumps, now non-existent, or garden dumps. However, 34 species of the 91 that he considered "fugitive" have become well established in the county, such as **Lapsana communis, Epipactis helleborine** and **Hesperis matrionalis**, and are placed in the main catalogue of this work. Several species found in Hoffmann's main catalogue have been judged to be non-persistent by the author. Most recently discovered aliens have been placed in the main catalogue since they appear to be colonizing disturbed habitats which are widespread. But many of these new alien species may not persist very long in our flora either.

Aconitum napellus L. Garden-monkshood
Agrostemma githago L. Corn-cockle
Amaranthus deflexus L. Low Amaranth
Amaranthus palmeri Wats. Palmer's Amaranth
Amaranthus powellii Wats. Powell's Amaranth
Amaranthus spinosus L. Spiny Amaranth
Amaranthus tuberculatus (Moq. ex DC.) Sauer Water Hemp
Brassica oleracea L. Cabbage
Calystegia pubescens Lindl. Hairy Bindweed
Chelone lyonii Pursh Pink Turtlehead
Chenopodium bonus-henricus L. Good-King-Henry
Chenopodium urbicum L. City Goosefoot
Consolida ambigua (L.) Ball & Heyw.
Coreopsis tinctoria Nutt. Garden Tickseed
Coreopsis lanceolata L. Lanceleaf Tickseed
Cucumis melo L. Melon
Cucurbita maxima Dcne. Squash
Fagopyrum tataricum (L.) Gaertn. India Buckwheat
Fumaria officinalis L. Fumitory
Glycine max (L.) Merr. Soybean
Helianthus X **kellermanii** Britt. Kellerman's Sunflower
Helianthus giganteus L. X **H. grosseserratus** Martens
Helianthus grosseserratus Martens Sawtooth Sunflower
Helianthus X **laetiflorus** Pers. Showy Sunflower
Ipomea purpurea (L.) Roth Purple Morning-glory
Lilium lancifolium Thunb. Tiger Lily
Lychnis chalcedonica L. Scarlet Lychnis
Lycopersicon esculentum Mill. Tomato
Marrubium vulgare L. Horehound
Medicago polymorpha L. Bur-clover
Medicago arabica (L.) Huds. Spotted Medick
Medicago minima (L.) Bartal. Little Medick
Monarda media Willd. Purple Bergamot
Neslia paniculata (L.) Desv. Ball Mustard
Nicandra physalodes (L.) Gaertn. Apple-of-Peru
Nicotiana alata Link & Otto Winged Tobacco
Papaver rhoeas L. Field Poppy
Petunia violacea Lindl. Garden Petunia
Philadelphus caucasicus Koehne Mock-orange
Phlox maculata L. Wild Sweet William
Prunus persica (L.) Batsch Peach
Ratibida pinnata (Vent.) Barnh. Greyhead Prairie Coneflower
Reseda odorata L. Garden Mignonette
Reseda lutea L. Yellow Mignonette
Rosa spinossima L Scotch Rose
Rudbeckia subtomentosa Pursh Sweet Coneflower
Salsola iberica Sennen and Pau Saltwort
Saponaria vaccaria Medic. Cow Herb
Scabiosa ochroleuca L. Yellow Scabious
Silene armeria L. Catchfly
Silene dichotoma Ehrh. Forking Catchfly
Silene dioica (L.) Clairville Red Campion
Symphytum asperum Lepechin Rough Comfrey
Triticum aestivum L. Wheat
Ulmus procera Salisb. English Elm
Xanthium spinosum L. Spiny Clotbur

Excluded Species

The following taxa, which have been listed for Berkshire County, are excluded because no specimen has been found, the specimen is inadequate for identification or the specimen has been misidentified.

Agastache nepetoides (L.) Kuntze no specimen
Aster foliaceus Lindl. misidentified
Aster lowrianus Porter inadequate specimen
Aster tradescantii L. misidentified
Crataegus crus-galli L. no specimen
Eragrostis minor Host. misidentified
Geum virginianum L. no specimen
Glyceria X **laxa** Scribn. no specimen

Lechea villosa Ell. misidentified
Lechea maritima Leggett ex BSP. inadequate specimen
Muhlenbergia sobolifera (Muhl.) Trin. no specimen
Physalis pubescens L. var. **integrifolia** (Dunal) Waterfall
no specimen
Potamogeton longiligulatus Fern. (**P. strictifolius** Benn.)
no specimen
Rosa virginiana Mill. inadequate specimen
Sagittaria rigida Pursh inadequate specimen, no
inflorescence
Taraxacum palustre (Lyons) Symons misidentified
(See Brunton, 1989)
Tradescantia virginiana L. no specimen

Table 3. Summary and comparison of 1995 flora with Hoffmann's 1922 flora. The 1995 flora includes all separate entries: nominate, variety and recognized hybrids.These figures are difficult to compare since elimination of many of Hoffmann's varieties, merging of taxa, and exclusion of certain species are not taken into account. Hoffmann listed 209 varieties, while 64 are noted in 1995.

	1995	1922
Total native plant species, including varieties	1222	1315
Total alien plant species, including varieties	453	271
Total plant species, varieties and hybrids	1675	1586
Percentage of flora that is alien	27%	17%
Native species new to county since 1922	35	
Native species new to state since 1922	12	
Alien species new to county since 1922	107	

References Cited

Andrews, A. L. 1900. Orchids of Mt. Greylock, Massachusetts. Rhodora 2:179-180.

Andrews, A. L. 1901. Several uncommon fern-allies from northwestern Massachusetts. Rhodora 3:252-253.

Andrus, R. E. 1980. Sphagnaceae (Peat Moss Family) of New York State. New York State Museum Bull. No. 442. Albany, NY.

Annual Reports. 1900-1979. City Engineer's Report. City of North Adams, MA.

Argus, G. W. 1986. The genus Salix (Salicaceae) in the Southeastern United States. Syst. Bot. Monogr. 9: 1-169.

Arnold, H. J. and S. W. Bailey. 1965. Bartholomew's Cobble. The Trustees of Reservations. Milton, MA.

Ballard, H. H. 1897. Amos Eaton, a pioneer of science in Berkshire County. Collections of the Berkshire Historical and Scientific Societies. Vol. 2:186-234.

Barrington, D. S. 1986. The morphology and cytology of *Polystichum* X*potteri* hybr. nov. (= *P. acrostichoides* X *P. braunii*). Rhodora 88:297-313.

Bascom, J. 1899. Appearance and disappearance of plants. Collections of the Berkshire Historical and Scientific Societies. Vol. 3:301-315.

Bierman, P. R. and D. P. Dethier. 1986. Lake Bascom and the deglaciation of Northwestern Massachusetts. Northeastern Geology 8:32-43.

Bormann, F. H. and M. F. Buell. 1964. Old-age stand of hemlock-northern hardwood forest in central Vermont. Bull. Torrey Bot. Club 91:451-465.

Braun, E. L. 1950. Deciduous forests of eastern North America. Blakiston Co. Philadelphia.

Bromley, S. W. 1935. The original forest types of New England. Ecol. Monogr. 5:63-89.

Brooks, R. R. 1987. Serpentine and its vegetation. Dioscorides Press, Portland, OR.

Brooks, R. R. R. 1953. Williamstown, the First Two Hundred Years. The McClelland Press. Williamstown, MA.

Brunton, D. F. 1989. The marsh dandelion (*Taraxacum* Section Palustria; Asteraceae) in Canada and the adjacent United States. Rhodora 91:213-219.

Bulkeley, M. 1969. A Berkshire tree list. The Berkshire Eagle, Pittsfield, MA. May 22.

Burk, C. J. and V. Prabhu. 1988. Growth and extension of a naturalized western Massachusetts stand of *Catalpa speciosa* Warder. Rhodora 90:457-459.

Carroll, D. 1993. Old growth forests in Massachusetts. Natural Heritage News 3(1):3.

Case, F. W., Jr. 1987. Orchids of the Western Great Lakes Region. Bull. Cranbrook Inst. Sci. 48.

Chadbourne, P. H. 1858. Natural history catalogue for Williamstown, Massachusetts. Williams Quarterly 5:342-359.

Chaney, R. W. 1947. Tertiary centers and migration routes. Ecol. Monogr. 17:141-148.

Churchill, J. R. 1899. Some plants about Williamstown. Rhodora 2:24-26.

Coddington, J. A. and K. G. Field. 1978. Rare and endangered vascular plant species in Massachusett. The New England Botanical Club in cooperation with The U. S. Fish and Wildlife Service. [Newton Corner, MA].

Cogbill, C. V. 1987. The boreal forests of New England. Wild Flower Notes. New England Wild Flower Society 2:27-36.

Conuel, T. 1994. The evangelist of old growth. Mass. Audubon Soc. Sanctuary 33(3):5-8.

Crane, J. W. 1918. History of the Town of Washington, Massachusetts. Typescript. Berkshire Athenaeum, Pittsfield, MA.

Cronon, W. 1983. Changes in the Land. Hill and Wang. New York, NY.

Cronquist, A. 1981. An Integrated System of Classification of Flowering Plants. Columbia Univ. Press. New York, NY.

Cronquist, A. 1982. Map of the floristic provinces of North America. Brittonia 34:144-145.

Crow, G. E. and C. B. Hellquist. 1983. Aquatic vascular plants of New England: Part 6. Trapaceae, Haloragaceae, Hippuridaceae. N. H. Agric. Exp. Sta. Bull. 524.

Dann, K. T. 1988. Traces on the Appalachians. Rutgers Univ. Press. New Brunswick, NJ.

Dansereau, P. and F. Segadas-Vianna. 1952. Ecological study of the peat bogs of eastern North America. Can. J. Bot. 30:490-520.

Davis, A. N. and T. L. Briggs. 1986. Dispersion pattern of aerial shoots of the common marsh reed *Phragmites australis* (Poaceae). Rhodora 88:325-330.

Davis, M. B. 1965. Phytogeography and palynology of Northeastern United States. pp. 377-401. In H. E. Wright and D. Frey, Eds., The Quaternary of the United States. Princeton Univ. Press. Princeton.

Davis, M. B. 1983. Quaternary history of deciduous forests of eastern North America and Europe. Ann. Missouri Bot. Gard. 70:550-563.

Davis, M. B. 1986. Climatic instability, time lags, and community disequilibrium. pp. 269-284 in Community Ecology. J. Diamond and T. J. Case, Eds. Harper and Row. New York, NY.

Davis, M. B. and J. C. Goodlett. 1960. Comparison of the present vegetation with pollen-spectra in surface samples from Brownington Pond, Vermont. Ecology 41:346-357.

Davis, R. B. and G. L. Jacobson Jr. 1985. Late glacial and early Holocene landscapes in northern New England and adjacent areas of Canada. Quaternary Res. 23:341-368.

Department of Environmental Management, Region 5, Commonwealth of Massachusetts. Undated. Unpublished materials for GOALS Plan for Savoy State Forest.

Dewey, C. 1818. Catalogue of the plants of Williamstown. Manuscript. Williamsiana, Williams College, Williamstown.

Dewey, C. 1829. Catalogue of plants. pp. 43-86. In D. D. Field, A History of the County of Berkshire by gentlemen of the county. Samuel Bush. Pittsfield, MA.

Dewey, C. 1840. Report on the Herbaceous Plants of Massachusetts. Commissioners of the Zoological and Botanical Survey of the State, Cambridge.

Dickinson, C. H. and G. J. F. Pugh. 1974. Biology of Plant Litter Decomposition, Vol. 2. Academic Press. New York.

Dore, W. G. and J. McNeill. 1980. Grasses of Ontario. Research Branch Agriculture Canada. Monograph 26.

Drew, B. A. 1982. Hard Times. pp. 3-5 in Berkshire off the Trail. Attic Revivals Press. Great Barrington, MA.

Dunwiddie, P. W. 1993. Survey of old growth forest in Massachusetts. MA Division of Fisheries and Wildlife, Natural Heritage and Endangered Species Program.

Dwight, T. 1822. Travels; in New-England and New-York, III. Timothy Dwight, New Haven.

Eaton, A. 1818. Manual of Botany for the Northern and Middle States, 2nd ed. Webster and Skinners. Albany, NY.

Egler, F. E. 1936. Ecological interrelationships and phytogeographical affinities of the Plateau vegetation. Ph. D. Thesis. Yale Univ., New Haven.

Egler, F. E. 1940. Berkshire Plateau vegetation, Massachusetts. Ecol. Monogr. 10:147-192.

Egremont Bicentennial Committee. 1976. Egremont 1756-1976 A History. The Berkshire Courier. Great Barrington, MA.

Federal Writer's Project. 1938. The Berkshire Hills. Duell, Sloan and Pearce. New York, NY.

Fernald, M. L. 1942. Some early botanists of the American Philosophical Society. Proc. Amer. Phil. Soc. 86:63-71.

Fernald, M. L. 1950. Gray's Manual of Botany. 8th ed. American Book Co. New York, NY.

Fernald, M. L., R. C. Bean and C. H. Knowlton. 1919. Plans for spring field trip of the New England Botanical Club. Rhodora 21:86-88.

Field, D. D. 1829. A History of the County of Berkshire by gentlemen of the county. Samuel Bush. Pittsfield, MA.

Fisher, J. T. 1933. New England's Forests: Biological factors. pp. 213-223 in New England's Prospects: 1933. J. W. Wright, Ed. Amer. Geogr. Soc. New York, NY.

Freckmann, R. W. 1978. New combinations in *Dichanthelium* (Poaceae). Phytologia 39:268-272.

Freckmann, R. W. 1981. Realignments in the *Dichanthelium acuminatum* complex (Poaceae). Phytologia 48:99-110.

Gleason, H. A. 1952. The New Britton & Brown Illustrated Flora of the Northeastern United States and Adjacent Canada. 3 Vols. New York Botanical Garden, New York, NY.

Gleason, H. A. and A. Cronquist. 1964. The Natural Geography of Plants. Columbia Univ. Press, New York.

Gould, F. W. and C. A. Clark. 1978. *Dichanthelium* (Poaceae) in the United States and Canada. Ann. Missouri Bot. Gard. 65:1088-1132.

Grace, J. B. and R. G. Wetzel. 1981. Habitat partitioning in cattails (*Typha*): experimental field studies. Amer. Nat. 118:463-474.

Graham, A. 1993. History of the vegetation. In Flora of North America. Vol. 1. Oxford Univ. Press, NY.

Grear, J. W. 1966. Cytogeography of *Orontium aquaticum* (Araceae). Rhodora 68:25-34.

Gustafson, A. H. 1937. A note on Amos Eaton's herbarium. Rhodora 39:153-155.

Hardin, J. W. 1973. The enigmatic chokeberries (*Aronia*, Rosaceae). Bull. Torrey Bot. Club. 100:178-184.

Hellquist, C. B. 1975. Correlation of selected dissolved substances and the distribution of *Potamogeton* in New England. Ph. D. Thesis. Univ. of New Hampshire, Durham, NH.

Hellquist, C. B. 1980. Correlation of alkalinity and the distribution of *Potamogeton* in New England. Rhodora 82:331-344.

Hellquist, C. B. 1984. Observations of *Potamogeton hillii* Morong in North America. Rhodora 86:101-111.

Henderson, R. 1982. Vegetation-fire ecology of Tallgrass prairie. Natural Areas J. 2:17-26.

Hermann, F. J. 1958. *Euphrasia micrantha* new to North America. Rhodora 60:307-308.

Hoffmann, R. 1899. Three plants of Berkshire County, Massachusetts. Rhodora 1:229-230.

Hoffmann, R. 1904. Notes on the flora of Berkshire County, Massachusetts. Rhodora 6:202-206.

Hoffmann, R. 1922. Flora of Berkshire County. Proc. Boston Soc. Nat. Hist. 36:171-382.

Holmes, G. W. and W. S. Newman. 1971. Surficial geologic map of the Ashley Falls Quadrangle, MA-CN. U. S. Geological Survey.

Hultén, E. 1970. The Circumpolar Plants. II. Almquist & Wiksell, Stockholm.

Johnson, C. W. 1985. Bogs of the Northeast. Univ. Press of New England. Hanover NH.

Jones, A. G. 1978. The taxonomy of *Aster* Section Multiflori (Asteraceae) I. Nomenclatural review and formal presentation of taxa. Rhodora 80:319-357.

Jones, A. G. 1980. Data on chromosome numbers in *Aster* (Asteraceae), with comments on the status and relationships of certain North American species. Brittonia 32:240-261.

Kadlec, J. A. 1958. An analysis of a woolgrass (*Scirpus cyperinus*) community in Wisconsin. Ecology 39:327-332.

Kartesz, J. T. 1994. A Synonymized Checklist of the Vascular Flora of the United States, Canada and Greenland. 2nd ed. Timber Press, Portland, OR.

Keith, H. F. 1912. History of Taconic and Mount Washington. Berkshire Courier Printers. Great Barrington, MA.

Knowlton, C. H. 1919. An excursion to Mt. Washington, Massachusetts, and Bash-bish Falls. Rhodora 21:198-202.

Kruckeberg, A. K. 1954. Plant species in relation to serpentine soils. Ecology 35:267-274.

Küchler, A. W. 1964. Potential Natural Vegetation of the conterminous United States. Amer. Geogr. Soc. Special Publ. 36. New York.

Kudish, M. 1979. Catskill soils and forest history. The Catskill Center for Conservation and Development. Hobart, NY.

Larson, J. A. 1980. The Boreal Ecosystem. Academic Press, New York.

Leak, W. B. 1978. Relationship of species and site index to habitat in the White Mountains of New Hampshire. USDA For. Serv. Res. Pap. NE-397.

Lewis, W. H. 1957. A Monograph of the Genus *Rosa* in North America east of the Rocky Mountains. Ph. D. Thesis. Univ. of Virginia, Charlottesville, VA.

Li, H-L. 1952. Floristic relationships between eastern Asia and eastern North America. Trans. Amer. Philos. Soc. 42:371-429.

Little, E. L. 1971. Atlas of United States Trees. Vol. 1. Conifers and Important Hardwoods. Misc. Publ. 1146. USDA Forest Service. Washington, D. C.

Manning, W. E. 1950. A key to the hickories north of Virginia with notes on the two pignuts, *Carya glabra* and *C. ovalis*. Rhodora 52: 188-199.

Manning, W. E. 1973. The northern limits of the distributions of hickories in New England. Rhodora 75:34-51.

Marchand, P. J. 1987. Life in the Cold. University of New England Press. Hanover, NH.

Massachusetts Historical Commission. 1994. Personal Communication.

McCann, J. A. and L. M. Daly. Undated. An inventory of the lakes, ponds and reservoirs of Massachusetts, Berkshire and Franklin counties. Publication No. 10-2. Water Resources Research Center, Univ. of Massachusetts, Amherst, MA.

McVaugh, R. 1958. Flora of the Columbia County Area, New York. N.Y. State Mus. Bull. No. 360.

Merrill, E. D. 1946. The Amos Eaton Herbarium. Rhodora 48:201-205.

Milham, W. I. Undated. Meteorology and Meteorological Observations in Williams College. The McClelland Press. Williamstown, MA.

Mitchell, R. S. 1986. A Checklist of New York State Plants. New York State Mus. Bull. No. 458.

Mitchell, R. S. 1988. Platanaceae through Myricaceae of New York State. New York State Mus. Bull. No. 464.

Moeller, R. W. 1984. The Ivory Pond project. North Amer. Archaeol. 5:1-5.

Moizuk, G. A. and R. B. Livingston. 1966. Ecology of red maple (*Acer rubrum* L.) in a Massachusetts upland bog. Ecology 47:942-950.

Monthly Traveller, The, or Spirit of the Periodical Press. 1838. An Expedition to Saddle Mountain. September 1838. Vol. 9(10). Boston, MA.

Motzkin, G. 1994. Calcareous fens of Western New England and adjacent New York State. Rhodora 96:44-68.

Natural Heritage and Endangered Species Program. 1994. Massachusetts List of Endangered, Threatened and Special Concern Species, Natural Heritage and Endangered Species Program, Boston.

Nature Conservancy News, The. 1989. The Nature Conservancy. Washington, D. C.

New England Weather Associates. 1900-1979. Unpublished weather data, North Adams, MA.

Nichols, G. E. 1916. The vegetation of Connecticut. V. Plant societies along rivers and streams. Bull. Torrey Bot. Club 42:235-264.

Niles, G. G. 1904. Bog-trotting for Orchids. G. P. Putnam's Sons. New York, NY.

Niles, G. G. 1905. Hoosac Valley and its flowers and ferns. Amer. Bot. (Binghamton) 9:1-7, 21-29.

Niles, G. G. 1913. The North American cypripediums. Amer. Bot. (Binghamton) 19:121-129.

Oltsch, F. M. 1974. The bogs of Berkshire County. M. S. Thesis. Univ. of Massachusetts, Amherst, MA.

Parrish, J. N., T. Marino and M. Bulkley. 1984. Recent paleontological discovery in Berkshire County. Bull. of the Massachusetts Archaeological Soc. 44:20-21.

Pielou, E. C. 1991. After the Ice Age; the Return of Life to Glaciated North America. Univ. of Chicago Press, Chicago, IL.

Potter, J. R., Jr. 1989. Observations of early season snow damage to mixed hardwood stands. A Journal 6:22-24. Center for Environmental Studies, Williams College, Williamstown, MA.

Ratcliffe, N. M. 1979. Field Guide to the Chatham and Greylock slices of the Taconic allochthon in western Massachusetts and their relationship to the Hoosac-Rowe sequence. pp. 388-404 in New England Intercollegiate Conference, 71st meeting.

Rawinski, T. 1983. Classification of natural communities in New England. The Nature Conservancy. Photocopy. Boston, MA.

Rehder, A. 1986. Manual of Cultivated Trees and Shrubs, 2nd ed. Dioscorides Press. Portland, OR.

Reschke, C. 1990. Ecological communities of New York State. N. Y. S. Dept. of Environmental Conservation. Latham, NY.

Reznicek, A. A. 1989. New England *Carex* (Cyperaceae): Taxonomic problems and phytogeographical considerations. Rhodora 91:144-153.

Reznicek, A. A. and P. W. Ball. 1980. The taxonomy of *Carex* Section Stellulatae in North America north of Mexico. Contr. Univ. Michigan Herb. 14:153-203.

Robbins, C. C. 1968. John Torrey (1796-1873), His life and times. Bull. Torrey Bot. Club 95:515-645.

Rodgers, A. D. 1942. John Torrey, A Story of American Botany. Princeton Univ. Press. Princeton, NJ.

Saterson, K. A. 1977. A vegetation history of Williamstown 1752-1977. Honors Thesis. Williams College, Williamstown, MA.

Schwintzer, C. R. 1981. Vegetation and nutrient status of northern Michigan bogs with a comparison to fens. Canad. J. Bot. 59:842-853.

Seischab, F. K. and J. M. Bernard.1991. Pitch pine (*Pinus rigida* Mill.) communities in central and western New York. Bull. Torrey Bot. Club. 118:412-423.

Seymour, F. C. 1982. The Flora of New England, 2nd ed. Phytologia Memoirs V.

Siccama, T. G. 1974. Vegetation, soil, and climate on the Green Mountains of Vermont. Ecol. Monogr. 44:325-349.

Siccama, T. G., F. H. Bormann and G. E. Likens. 1970. The Hubbard Brook ecosystem study: Productivity, nutrients and phytosociology of the herbaceous layer. Ecol. Monogr. 40:389-402.

Sorrie, B. A. 1987. Notes on the rare flora of Massachusetts. Rhodora 89:113-196.

Sorrie, B. A. 1989. Massachusetts flora: a review of current distribution and conservation of rare species. Rhodora 91:116-120.

Stanley, R. and N. M. Ratcliffe. 1985. Tectonic synthesis of the Taconian Orogeny in western New England. Geol. Soc. Amer. Bull. 96:1227-1250.

Swanton, H. 1981. Ralph Hoffmann: unsung guide to the birds. Natural History 90(11):20-30.

Takhtajan, A. 1986. Floristic Provinces of the World. University of Calif. Press. Berkeley, CA.

Technical Planning Assoc. 1959. Berkshire County, Massachusetts, A Regional Planning Study, Vol. 2. New Haven, CT.

Tryon, R. M. and A. F. Tryon. 1982. Ferns and allied plants with special reference to Tropical America. Springer-Verlag. New York, NY.

University of Massachusetts. 1992. Cooperative Extension System. Current issues in forestry. 4(1).

U. S. Dept. of Agriculture (U.S.D.A.), Soil Conservation Service. 1977. Water & Related Land Resources of the Berkshire Region. [Amherst, MA].

U. S. Dept. of Agriculture (U.S.D.A.), Soil Conservation Service. 1988. Soil Survey of Berkshire County, Massachusetts. [Amherst, MA].

U. S. Dept. of Commerce (U.S.D.C.). 1960-1969. U. S. Weather Bureau. Climatological Data.

U. S. Dept. of Commerce (U.S.D.C.). 1981-1987. U. S. Weather Bureau. Climatological Data.

Van Diver, B. B. 1987. Roadside Geology of Vermont and New Hampshire. Mountain Press Publ. Missoula, MT.

Voss, E. G. 1980. Michigan Flora. Part 1. Gymnosperms and monocots. 2nd ed. Bull. Cranbrook Inst. Sci. 55.

Voss, E. G. 1985. Michigan Flora. Part II. Dicots (Saururaceae-Cornaceae). Bull. Cranbrook Inst. Sci. 59.

Walker, R. B. 1954. Factors affecting plant growth on serpentine soils. Ecology 35:259-266.

Weatherby, C. A. 1947. Bartholomew's Cobble. Amer. Fern J. 37:1-6.

Webber, J. M. and P. W. Ball. 1984. The taxonomy of the *Carex rosea* group (section Phaestoglochin) in Canada. Canad. J. Bot. 62:2058-2073.

Whitehead, D. R. 1979. Late-glacial and postglacial vegetational history of the Berkshires, Western Massachusetts. Quaternary Res. 12:333-357.

Whipple, A. B. 1892. The early botany of Berkshire. pp. 5-35 in Berkshire book, by its Historical and Scientific Society. Sun Printing Co. Pittsfield, MA.

Wood, C. E. 1970. The distributional history of the biota of the Southern Appalachians, Part II: Flora Symposium Research Div. Monograph 2. Virginia Polytechnic Inst. and State Univ. Blacksburg, VA.

Woodward, F. I. 1987. Climate and Plant Distribution. Cambridge University Press. Cambridge, England.

Zen, E. (Ed.) 1983. Bedrock geologic map of Massachusetts. Dept. of Public Works. Commonwealth of Mass. U. S. Geologic Survey.

Zika, P. F. 1990. Range expansions of some grasses in Vermont. Rhodora 92:80-89.

Zika, P. F. and K. T. Dann. 1985. Rare plants on ultramafic soils in Vermont. Rhodora 87:293-304.

Useful field guides and local guides

Gleason, H. A., and A. Cronquist. 1991. Manual of Vascular Plants of the Northeastern United States and Adjacent Canada. 2nd. ed. New York Botanical Garden, Bronx, NY.

Laubauch, R. 1992. A Guide to Natural Places in the Berkshire Hills. Berkshire House. Stockbridge, MA.

Lellinger, D. B. 1985. A Field Manual of the Ferns & Fern-allies of the United States & Canada. 1985. Smithsonian Institution Press. Washington, DC.

Luer, C. A. 1975. The Native Orchids of the United States and Canada excluding Florida. New York Botanical Garden, Bronx, NY.

Newcomb, L. 1977. Newcomb's Wildflower Guide. Little, Brown and Co. Boston, MA.

Strauch, J. G. Jr. 1995. Wildflowers of the Berkshire & Taconic Hills. Berkshire House. Stockbridge, MA.

Index

A

B

Index

C

Index

D

Index

Index

G

H

Index

Index

M

N

Index

Index

Index

T

Index

U

V

W

Index

X

Y

Z